LAKE SCIENCES AND CLIMATE CHANGE

Edited by **M.Nageeb Rashed**

Lake Sciences and Climate Change

http://dx.doi.org/10.5772/61895
Edited by M.Nageeb Rashed

Contributors

Noha Donia, Ali Assani, Brian Kesner, Paul Marsh, Jamie Wisenall, Anatoliy Filonov, Rutger De Wit, Marella Thomas Kiran, Bhaskar Mallimadugula, Archana Tiwari, Eugenia López López, Jacinto Elías Sedeño-Díaz, Ricardo A. Ruiz-Picos, Meral Apaydın Yağcı, Alexis Joseph Rodríguez Romero, Erick Mendoza Martínez, Eugenia López-López

CBS, Edition **2017**

Published by InTech

Janeza Trdine 9, 51000 Rijeka, Croatia

Publishing Process Manager Romina Rovan

Technical Editor SPi Global

Cover Designer InTech Design Team

Additional hard copies can be obtained from orders@intechopen.com

Lake Sciences and Climate Change, Edited by M.Nageeb Rashed
p. cm.
Print ISBN 978-953-51-2556-3
Online ISBN 978-953-51-2557-0

INTECH

open science | open minds

Contents

Preface

Lakes are naturally or artificially created, while the majority of lakes on the earth are fresh water lakes. Lakes are highly essential for their water quality, habitat suitability, and water supply. Lakes constitute important sources for aquatic life, fish, wildlife, and human. Lakes can undergo rapid environmental changes, and any alteration in their water renewal rates and environmental quality has ecological and societal implications. One of the most common lake problems is eutrophication, which is the process of physical, chemical, and biological changes in the lake. Eutrophic conditions can be exhibited with algal blooms, sedimentation, turbidity, oxygen depletion, growth of aquatic plants, water-level changes, and shift of populations of plant and animal species.

Most important factors that affect lake ecosystem are climate (temperature, precipitation, wind, and solar radiation), atmospheric inputs, land use, morphology, and physiography. Several lake environmental problems can affect their water quality and habitat suitability. Climate around the world has been changing since 1880. The global temperature has risen by 0.85º C since 1880 and the surface warming amounting 3.7º C will be likely between 2081 and 2100. Lakes are always subject to the impacts of climate change. Monitoring the effects of climate change on lake ecosystem has been considered by scientists, and focus on research demonstrates the sensitivity of lakes to climate including physical, chemical, and biological properties.

The book " Lake sciences and climate change " deals with several aspects of lake sciences (botany, zoology, geology, chemistry, models, morphology, and physiography), as well as the effect of climatic changes on lake body ecosystem. The book is divided into three sections and structured into 10 chapters. The first section "Lake and Climatic Change" represents three chapters:

Chapter 1 overviews on climatic change in a large shallow tropical Lake Chapala, Mexico. Chapter 2 deals with the relationship between water levels in the North American Great Lakes and climate indices. Chapter 3 shows the Lake La Salada de Chiprana (NE Spain) as an example of an thalassic salt lake in a cultural landscape. The second section "Lake Biology and Health" discusses three chapters in detail: Chapter 1 assesses health condition with oxidative stress biomarkers in sentinel organisms in shallow lakes of the Mexican Central Plateau. Chapter 2 discusses variations in the zooplankton species structure of an eutrophic Lake. Chapter 3 discusses phycoremediation of eutrophic lakes using diatom algae. The third section "Water quality, management and modeling" contains four chapters: Chapter 1 discusses adaptive management of an imperiled catostomid in Lake Mohave, Lower Colorado River, USA. Chapter 2 discusses chemometric analysis of wetlands remnants of the former Texcoco Lake, a multivariate approach. Chapter 3 describes fuzzy logic as a tool to assess water quality for reservoirs, a regional perspective (Lerma river basin, Mexico). Chapter 4 deals with water quality modeling of Northern Lakes case study (Egyptian Northern Lakes).

Scientists from different fields of lake sciences reported in this book their findings. The book "Lake Sciences and Climate Change" offers an important information source for researchers and professionals working in lakes and climatic changes.

I would like to express my sincere thanks to Prof. Dr. M.G.Sheded, Prof. Dr. A.ElAutafy (Botany Department), and Dr. A.A.ElSadek, (Zoology Department), Faculty of Science, Aswan University, Egypt, as well as to Dr. Chenlin Hu, College of Public Health, Ohio State University, USA, for their contribution in the scientific revision of the book chapters.

M.Nageeb Rashed
Faculty of Science
Aswan University,
Egypt

Section 1

Lake and Climatic Change

Chapter 1

Climatic Change in a Large Shallow Tropical Lake Chapala, Mexico

Filonov Anatoliy, Iryna Tereshchenko,
Cesar Monzon, David Avalos-Cueva and
Diego Pantoja-González

Additional information is available at the end of the chapter

http://dx.doi.org/10.5772/63715

Abstract

Measurements of temperature, currents and lake level taken in 2005–2014 are analyzed and discussed. We obtained a conceptually new data set on the formation of the thermocline in Lake Chapala. It is showed that the thermocline in the lake occurs only during the daytime, in the top 0.5–1.0-m layer of the water column, whereby the vertical temperature gradient reaches 2.5°C/m within that layer. At night, the top layer is cooled, which causes strong vertical mixing down to the bottom. Moored measurements of temperature and level from Lake Chapala reveal the presence of seiches oscillations with periods of 5.7 and 2.8 hours with amplitudes of 15.4 and 8.1 mm. Temperature measurements on sections across the lake showed that in the northern part of the lake, the water column is warmer that in southern 2–3°C in all seasons. The lake currents were simulated for wet and dry seasons. The model results are in good agreement with the acoustic Doppler current profiler (ADCP) data. The presence of an anticyclone gyre in the central part of the lake in both seasons is detected.

Keywords: Lake Chapala, level, temperature and current measurements, hydrodynamic modeling, breeze

1. Introduction

Lake Chapala, the largest in Mexico and third largest in Latin America, has an average length and width of 75 × 22 km and an average depth of only 6 m, with a maximum of 11 m (**Figure 1**). Among shallow lakes, Lake Chapala is the largest in the world [1–3]. It plays an important role in the economy of the region. Approximately 20 million people live in the vicinity of Lake Chapala. The area also includes important industrial and cultural sites.

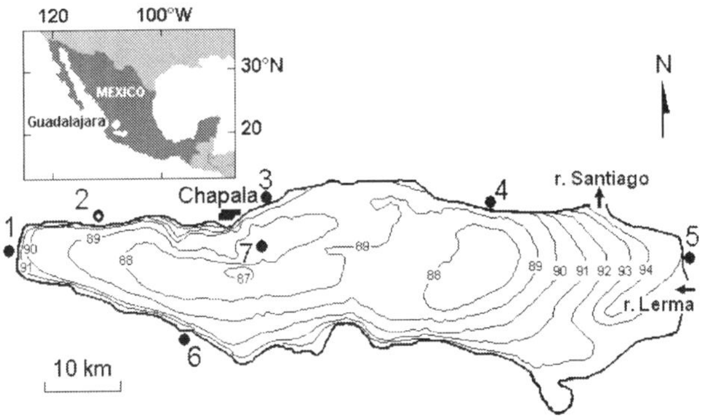

Figure 1. Map of the Lake Chapala. The numbers refer to weather stations: 1. Jocotepec, 2. Atequiza, 3. Chapala, 4. Poncitlán, 5. Jamay, 6. Tizapán and 7. Isla de Alacranes.

The lake provides for a mild climate and establishes a land-lake breeze circulation throughout its costal area. Atmospheric humidity in this area is also moderate, which together with the pleasant landscape, makes the lake a great tourist attraction [2, 3].

The lake's tributary and outflow rivers are the Lerma and Santiago, respectively. Both rivers and the lake form a unique reservoir system covering an area of approximately 47,000 km^2. The annual rainfall is 750 mm, and evaporation from the lake surface ranges from 1000 to 1400 mm per year, resulting in a negative balance [2, 3]. The deficit is offset by inflow from the Lerma River [4–6] in dry years annual rainfall can drop to 500 mm and the lake does not contribute any water to the Santiago River [2, 3, 7]. Tereshchenko et al. [2] mentions that in the high-rainfall years, the annual precipitation can reach 1000 mm and an important volume of water exits through this same river.

The Lerma River is the main tributary of Chapala Lake. Tereshchenko et al. [2] mentions that according to data from the National Water Commission (Comisión Nacional del Agua, México; CNA), approximately 2.75×10^6 m^3 [2, 3, 8] of suspended particles enter the lake via its tributary from the lake's watershed. The size of the particles can reach up to 0.5 μm and are mainly deposited in the eastern part of the lake. The result is a decreasing depth by the accumulation of particles in the bottom of the lake. Suspended particles modify the water transparency.

The principal loss of water from the lake is through evaporation. Tereschenko et al. [2] mentions that during the spring months, from March to May, evaporation brings down the lake water level, on average, 10 mm a day. The shallow depths (1–2 m) and low transparency of water in the east part of the lake create less thermal inertia and allow for stronger heating in comparison with deeper and more transparent portions of the lake, resulting in higher evaporation rates [9]. In this connection, Filonov [8] proposed the construction of a dam to cut off the eastern 20% of the lake to lower these losses. Detailed measurements show [5] that the shallow east sector has a surface water temperature of as much as 3°C above that in the central area of the lake. However, on the basis of these measurements, it is difficult to advance any conclusion about the east sector as being the main evaporation area of the lake. Resolution of this controversy would require a surface water temperature-monitoring program (weekly measurements) of year duration, at least.

Sandoval [1], Tereschenko et al. [2] and Avalos-Cueva et al. [3] mention that the lake has reached its current status due to water usage of the Lerma for irrigation and industrial needs and to cover the domestic demand of Guadalajara city. Also, dewatering for agricultural irrigation and neighboring villages of the lake has a negative balance in storage. Therefore, Sandoval [1] states that the lake will be dry out in a future. However, not all researchers agree with this statement. Filonov [8] puts forward some different arguments based on numerical modeling. He states that the main cause of the long-period-level fluctuation is unfavorable climatic factors and not just anthropogenic influence. But he fails to definitively establish the causes of Chapala Lake water-level fluctuation, as well as of rainfall fluctuations over its watershed.

Filonov [8] also shows that the hydrometeorological regimen of Chapala Lake is influenced by El Niño episodes. The El Niño 1997–1998 event was the strongest of this century [10, 2]. This event registered unusually high air temperatures and consequently a very intense dry season in the west-central Mexico. In Guadalajara, Jalisco the monthly average air temperature from March to May 1998 was 3–4°C higher than the climatological mean for this same season. The atmospheric relative afternoon humidity decreased from 6 to 8% [6]. In the dry season, the lake lost reaches up to 1 m due to evaporation water demand [8].

Studies on the quality of lake water and pollutant dispersion are pretty few and discrete points, while the thermodynamic conditions have not been studied in detail. On such a situation is not that simple to propose strategies to rescue the lake. A modern method for monitoring the thermal regime of epicontinental water bodies has been made through the use of satellite images. Such methods found wide application in physical oceanography [11] and physical limnology [2, 12].

In the past two decades, researchers from the Physical Department of the University of Guadalajara began to study the thermodynamic processes in the Lake Chapala using hydrodynamic modeling and the analysis of data collected with the use of up-to-date oceanographic and meteorological measurement devices. In this work, we discuss the analysis of wind data collected over the lake, as well as the fluctuations in water temperature, currents and lake level. The main purpose of this study is to gain more understanding of the thermal and dynamic patterns of the Lake Chapala during the dry and wet seasons.

2. Measurements and data sets

One major impediment in the data collection across the study area is a great number of fishing nets deployed in the Lake Chapala. Hundreds of people are engaged in the fishing industry, which is a major source of income for them. Therefore, working there, we always rely on good luck, but do not always succeed. On some occasions, partial losses of the instruments and equipment were inevitable.

This study is based on the analysis of the temperature, currents and lake-level data collected in 2005–2014 using the following oceanographic instruments: CTD SBE19-plus, SBE-39, SBE-26, HOBO V2 and a ADCP RDI 600 kHz, ADP SONTEK 1000 kHz. During that time period, the measurements were not taken regularly as they pursued different goals. The sampling strategies varied with the experiment (we used different sets of instruments) and will be described in the corresponding sections.

Most measurements were taken in the deeper northern part of the lake. The meteorological data were collected from the network of seven automatic meteorological stations deployed around the lake and in its center. The spatial structure of the temperature field and currents for the dry and wet seasons were sampled by towed temperature recorders arranged in the antenna pattern and ADCP.

3. Results

3.1. Long-term fluctuations in the lake level

In last century, Lake Chapala has suffered two catastrophic declines of its level. From 1945 to 1955, the level fell by nearly 4 m and then in the next 4 years was increased by 5m. In 1977, again a decrease began and by 1989 was down nearly 5 m. Then, for the year 1995, the level rose by 2 m and currently undergoes a period lake-level decline, but still lack about 0.5 m to reach the level of 91.9 m maximum reduction achieved in 1955 (**Figure 2**).

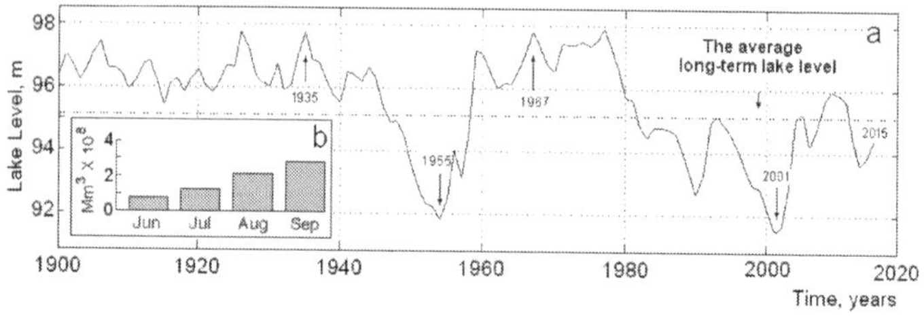

Figure 2. (a) Historical average annual water levels in Lake Chapala (Data from Water National Council of Mexico). Monthly mean Lerma River discharge in Chapala Lake (in millions cubic meters) during rain season from 2001. to 2007 (b).

The level oscillations with periods of 20–50 years are presented in all lakes in the world [7]. The cause of this is the climate oscillation in the territory of the basin, caused by the cyclical variation in solar activity and the succession of epochs of atmospheric circulation. Within each cycle phase, plenty of water (transgressive) is changed to that of low water (reverse). The increase or decreases in the water level in the lakes observed in these phases are the result of cyclical fluctuations in the intensity of rainfall.

The study of Filonov [8] also shows that the hydrometeorological regimen of Chapala Lake is influenced by El Niño episodes. The El Niño 1997–1998 event was the strongest of this century [10, 13]. This event registered unusually high air temperatures and consequently a very intense dry season in the west central of Mexico. In Guadalajara, Jalisco the monthly average air temperature from March to May 1998 was 3–4°C higher than the climatological mean for this same season. The atmospheric relative afternoon humidity decreased from 6 to 8% [14]. Due to the intense dry season, the lake lost more than 1 m of its water level due to evaporation and the excess of water pumped to satisfy water demand from the local industry and population [8].

3.2. Level fluctuations caused by free long gravity waves (seiches)

Cross analysis with wind and atmospheric pressure fluctuations has shown that sharply amplified seiches always after pressure increases that are caused by synoptic processes. A pulse of pressure acting on Lake Chapala obviously causes an inclination of its level on one of its parts (because of the large linear size of the lake) and then these pulses generate long free gravitational waves causing horizontal currents and level fluctuations [3, 9]. Undoubtedly, the lake breeze also causes fluctuations of lake level with daily periodicity, causing a "wind tide" on the coast in the afternoon and at its center in the night. The Lake acquires this energy, which eventually results in free lake-level fluctuation, with periods near of 6 hours and forced fluctuations with periods of 24 hours.

Basic equations describing seiches can be written in the following form [15]:

$$\frac{\partial u}{\partial t} - fv = -c^2\frac{\partial \eta}{\partial x} + \tau_x \tag{1}$$

$$\frac{\partial v}{\partial t} + fu = +c^2\frac{\partial \eta}{\partial y} + \tau_y \tag{2}$$

$$\frac{\partial u}{\partial x} + \frac{\partial v}{dy} + \frac{\partial \eta}{\partial t} = 0. \tag{3}$$

where x and y are the horizontal coordinates, u and v are the corresponding components of the vertically integrated current, t is a time, $\eta(x,y)$ is the free surface perturbation, f is the Coriolis parameter, $c=\sqrt{gH}$ is the speed of propagation of long gravity waves, g is the gravitation constant, H is the average depth of water, and τ_x,τ_y are the horizontal components of the force that causes seiches.

Solving Eqs. (1)–(3) can be accomplished only by numerical methods because their solution depends on the lake form. Let us make some estimates on the seiches without fully solving the given system of the equations. In long, narrow and shallow lakes ($H/L \ll 1$, L is the length of the lake), such Lake Chapala, it is possible to neglect geostrophic effects and vertical acceleration. Then, the solution for the current speed u (v - the cross component is very small), the free surface perturbation, and period of the seiches can be approximated under the following formulas [9, 16]:

$$u = u_{\max} \sin(n\pi x / L) \sin(2\pi t / \tau_s), \tag{4}$$

$$\eta = A_s \cos(n\pi x / L) \cos(2\pi t / \tau_s), \tag{5}$$

$$\tau_s = 2L / (n\sqrt{gH}). \tag{6}$$

u_{max} is the maximal current through a nodal line; A_s is the seiche amplitude at the edges of the lake; n is the number of seiches nodes. An Eq. (6) is known in the literature by the name of "Merian formula" [15].

Figure 3 shows an example of such oscillations; measured near the shore on the hydro meteorological station Chapala in 1997 (**Figure 1**). It was the first measurement of this kind on the lake. Let us estimate some parameters of the seiches in Lake Chapala, using the almost pure harmonic fluctuations (**Figure 3d**) having an average height (the double amplitude) of 15 mm and a period about 6.0 hours. These fluctuations should undoubtedly cause one-nodal seiches. Seiches period by the Eq. (6) gives a value of 5.8 hours, though this value is slightly less than the one described above by spectrum analysis. However, it can be used to calculate the data below.

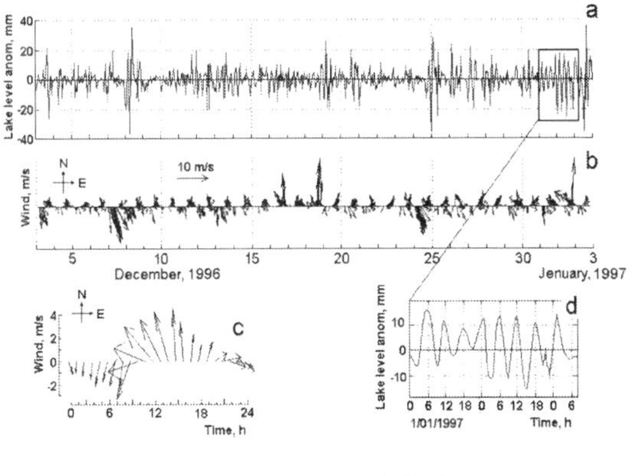

Figure 3. The fluctuations of the lake level anomalies (a). The panels (b) show the hourly course of the wind and daily average wind for 1. month (c). Almost pure sines fluctuations in lake level are described in (d).

3.3. Fluctuations in the level and the temperature at a fixed point of the lake

The main power source of the lake breeze circulation is the diurnal temperature cycle caused by the daily variations of (a) incoming solar radiation and (b) heating of the underlying surface and to atmosphere. The interaction between the land, lake and atmosphere is a very complex system with many feedbacks. The area around the Lake Chapala is mountainous, with valleys of various spatial orientations. The thermal energy pulsates with daily periodicity but does not remain at a fixed frequency. It is redistributed at different frequencies in a complex way in the form of fluctuations [17].

The principal dynamic process that occurs in the lake is the lake breeze circulation. Daytime breeze speed does not usually exceed 4 m/s. Beyond any doubt, the lake breeze causes the increase in the evaporation from the lake's surface. Lake Breeze, together with atmospheric pressure variations, generates free seiches waves.

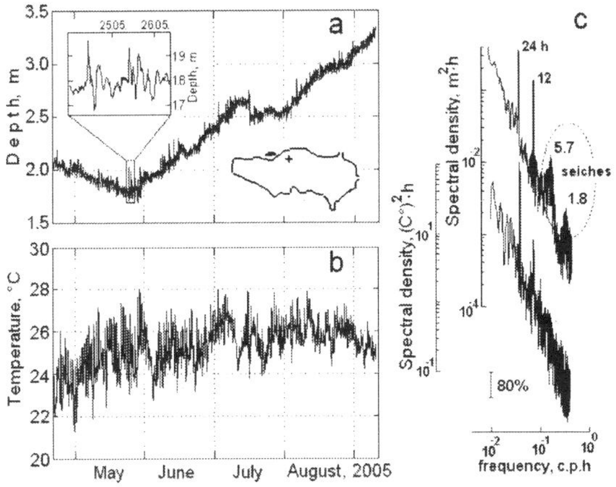

Figure 4. Hourly fluctuations of the lake level (a) and temperature fluctuations (b) at the mooring site during the summer 2005. experiment. Frequency spectra of the lake level and temperature fluctuations (c). The numbers designate the periods of the main peaks in the spectra. The vertical line designates the 80% confidence interval.

For a detailed study of the level and temperature variations at a fixed point of the lake, we made a special experiment. The measurements were taken from 20 April to 12 September 2005 (153 days). The lake level and temperature data were sampled every 5 min by a SBE-26 temperature-depth recorder (the accuracy is 0.002°C for temperature and 1 mm for depth). The device was deployed on the mooring in the northern part of the lake, at the water depth of 2 m. **Figure 4a** shows that the seasonal-level fluctuations in 2005 strongly depend on the evaporation and precipitation over the lake and its watershed. In April and May, the level dropped at a rate of about 30 cm/month, and then, until the middle of September, it raised more rapidly, about 50 cm/month. In July, the level remained almost unchanged because of the reduced rainfall during this time of the year, which usually occurs over the territory of Central Mexico and is called "canicula" [3, 10].

The SBE-26 time series also shows that the daily fluctuations of lake level are determined by diurnal and semidiurnal harmonics. At a single point, the level fluctuations occur at times of the amplification and easing of the breeze whose speed was measured at the weather station Chapala [18]. The breeze forcing is the main source of energy for all kinds of motion in the lake. With a well-defined diurnal cycle, this breeze varies from virtually calm at night and morning to steady northerly winds up to 6 m/s (gusting up to 10–12 m/s) in the afternoon. The lake-breeze circulation is important on the processes of vertical and horizontal mixing, since it is active throughout the year [6].

The analysis shows that the start of wind intensification lags the time of the downturn of the lake level by less than 2 hours. The rise and drop of the level are asymmetrical. The trough lasts longer than the peak, which is apparently caused by the asymmetric impact of the wind on the water surface.

As seen from the data (**Figure 4b**), the amplitude of the daily temperature fluctuations at the mooring site are 2–3°C at the 2 m level and decrease to 1–1.5°C as the instrument depth increases due to the higher lake level during the summer months.

Figure 4c shows the spectra computed from the time series of lake level and temperature variations. The level spectrum reveals the presence of free seiches waves with 5.7- and 2.8-hour period in the lake. Their mean square amplitudes are 15.4 and 8.1 mm, respectively. The lake is shaped like an ellipse whose axes vary in size; therefore, the oscillations with a period of 5.7 hours likely correspond to the seiches propagating along the greater axis of the ellipse (west-east). The other oscillation mode is related to the seiches propagating along its smaller axis (north-south).

We used the Merian equation (Eq. (6)) to evaluate the theoretical periods of the two principal waves, taking that the maximum length of the lake is 75 km, maximum width of 22 km and the average depth is 5 m [18]. Thus, the periods of the first and second modes of horizontal oscillation are 5.87 and 2.7 hours, respectively. These results closely match those obtained from our measurements. The seiches oscillation periods in Lake Chapala depend mainly on the mean depth of the lake, as the length along the major axis of the lake is almost unchanged. Interannual fluctuations of the Lake Chapala can reach 6 m, for example, in 1955–1960 [11], and therefore, in extreme years, the seiches periods may vary (increase or decrease) from 3 to 8 hours.

Our earlier study [18] reports almost similar results, except that there is no well-defined mode with the 2.8-hour period due to the weak amplitude. The seiches generate periodic currents, which peak at 1 cm/s in nodal line areas [18].

3.4. The diurnal variation of the lake temperature profile

Until now, no scientific work has addressed the issue of daily variations of water temperature in the Lake Chapala from in the entire water column. Many believe that in such a shallow lake the water temperature is uniform within a water column due to wind mixing [19]. Our research has shown that it is not true.

To study the variability of a vertical temperature profile, high-frequency measurements of water temperature in the entire water column were taken on 7–8 June 2005 from a boat anchored near the mooring. A boat-mounted meteorological station HOBO-logged wind speed and direction with a 10-min sampling rate. The water depth in this experiment was about 2 m. Water temperature was measured by a CTD 19-plus (Sea-Bird Electronics profiler) with temperature depth and accuracies of 0.005°C and 4 cm, respectively. The CTD profiler with a 0.25-s sampling rate was manually dropped and raised every 5 min with a speed of 0.1 m/s.

Chapala is a freshwater lake, and hence, the vertical change in water density is completely determined by temperature. To estimate the variability of the vertical stratification in the vicinity of the mooring, the buoyancy frequency was calculated: $N(z)=[(g/\rho)\cdot(\partial\rho/\partial z)]^{1/2}$, where z is the depth, $\rho=\rho(z)$ is the density of lake water, which was calculated under the formula: $\rho=\rho_o\,(1-1.96\times10^{-6}(T-289)(T-4^2)/(T+68.1)\times10^3)$, which is usually used for shallow lakes [20].

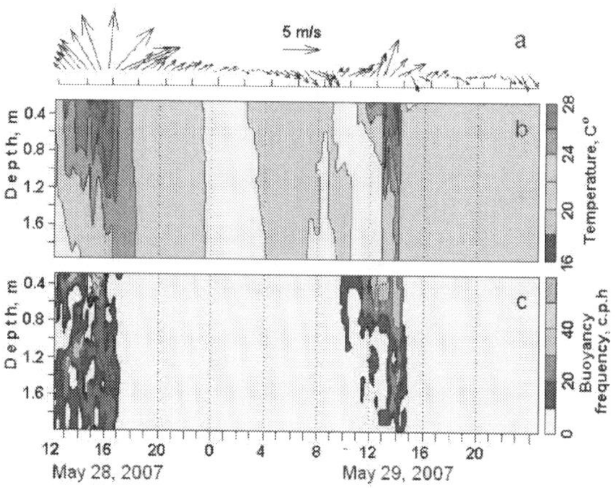

Figure 5. Variation of wind (a), temperature (b) and buoyancy frequency (c) in the vicinity of the mooring. The vertical casts were made every 10. min with a CTD-SBE-19 plus, which was manually dropped from an anchored boat with an up cast and down cast speed of about 0.1 m/s and a 0.25 s sampling rate.

The collected data sets are presented in **Figure 5**, which illustrates the connection between temperature fluctuations and strengthening (weakening) of the stratification caused by the breeze. The thermocline is formed around midday in the 1 m subsurface layer, in which the vertical temperature gradients reach 0.5°C/m and decrease to zero at the bottom. The cooling of the surface layer started at 20:00 and peaked at about 6:00, causing the cold water to sink to the bottom. Below the 0.4-m depth, the water temperature was almost homogeneous because of the intensive stirring. At daytime, the buoyancy frequency in the surface layer exceeded 50–60 c/hour (about a 1-min period).

The intensive subsurface thermocline serves as a "hard cap" and prevents the daytime heat and momentum exchange at the surface from penetrating into the deeper layers of the lake. Moreover, a similar stratification whereby a narrow surface waveguide for short internal wavelengths is generated was previously observed in the experiment in the eastern part of the lake [18]. In that experiment, it was shown that the short waves have periods of 10 min, amplitudes of 1 m and phase speeds of 0.15 m/s.

Thus, the daily variability of temperature stratification in the Lake Chapala prevents the water masses from mixing and contributes to the accumulation of heat in the surface layer during the daylight hours. Conversely, the cooling process results in negative vertical temperature gradients, which in turn cause the intense vertical flow. The latter powers the ascent of nutrients to the surface, which contributes to the high biological productivity of the lake.

3.5. North-south temperature cross section of the lake

The spatial distribution of surface temperature across the Lake Chapala was previously discussed in reference [2, 3], based on extensive satellite data sets with high spatial resolution. In that study, it was shown that the northern part of the lake is warmer than its central and southern parts, which is caused by the specific features of its circulation.

To confirm this finding and shed light on other processes that occur in the lake, north-south temperature cross sections were carried out in February, April, July and October 2006. Each survey included 60 equidistant bottom casts, 250 m apart. The SBE19-plus CTD profiler with 0.17-s sampling rate was manually dropped from the boat with a speed of about 0.1 m/s. The global position system (GPS) fixed the coordinates of the casts.

The measurements were taken in the morning (from 7:00 to 9:00 am), so that the temperature along the cross section was not biased. The cross section extends from the northern coast (near the town of Chapala) to the south side of the lake. The spatial distribution of temperature at the cross section is shown in **Figure 6**.

It is observed that the *in situ* measurements confirm the previous finding, namely that in all seasons the temperature in the central and northern parts of the lake is higher than in the southern part. This holds true not only for the surface but also for the bottom layer. In all four seasons, vertically averaged temperatures at the north and south ends of the cross section differ by 2–3°C. The vertical distributions of temperature on the sections are uneven.

The northern part of the section shows the penetration of warm water from the anticyclonic gyre which is stationary in the study area at morning time. In April, July and October, the water columns in the central and southern parts of the lake were poorly stratified, which was probably caused by vertical mixing at nighttime. These results shed new light on the thermal structure of the lake obtained and discussed in previous studies [2, 3, 18, 19, 21]. They should be taken into account in the design of future experiments in the study area and 3D model simulations.

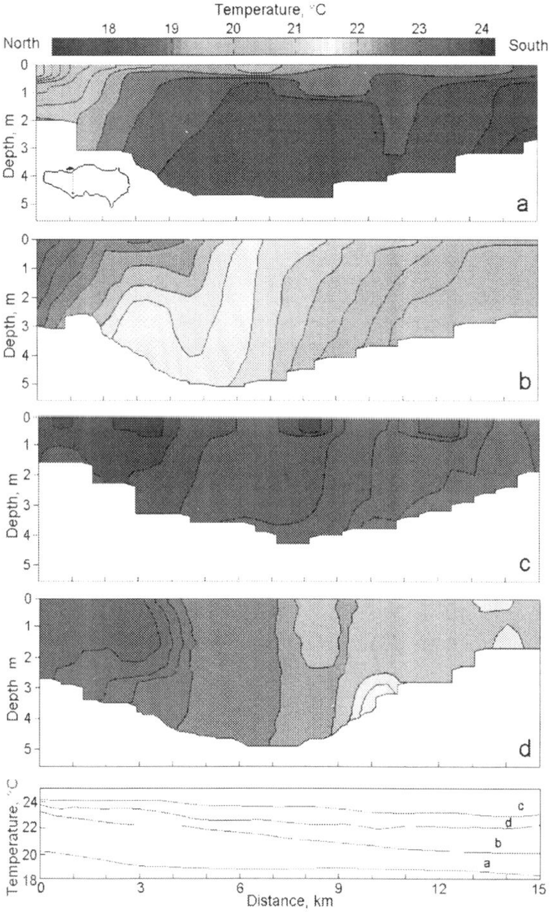

Figure 6. Vertical temperature cross sections in February (a), April (b), July (c) and October 2006. (d). The bottom panel shows the vertically averaged temperatures at the corresponding cross sections.

3.6. Current simulations in the lake for the dry and wet seasons

The results of numerical simulations of the Lake Chapala circulations are reported in some publications [3, 18, 21–26], but none of them is based on the experimental and observational wind data sets collected in the study area during different seasons. Therefore, we carry out this study.

The HAMSOM model (Hamburg Shelf Ocean Model) was used to model the horizontal currents resulting from wind circulation over the lake. The model is written in a finite difference scheme, using the Arakawa-C grid [27]. The HAMSOM model is vertically averaged, bi-dimensional, non-linear, semi-implicit, written in finite differences and represents the simultaneous solution to the Navier-Stokes and a continuity equations. The model was previously applied to the Gulf of California [28, 29]; the Santa Maria del Oro Volcanic Lake [30] and Lake Alchichica [31] and in the North Sea [32, 33].

To study the circulation patterns throughout the dry and wet periods, a simulation was conducted with the HAMSOM model forced by the wind field obtained from the network weather stations. The current simulation was carried out for the dry and wet season. The bathymetric grid had a 97 × 270 mesh: = Δy = 300 m and Δt = 30 s. The simulations started from a condition of no movement and were run for 30 days before obtaining a stable initial condition. The model was initialized with wind fields from the network of weather stations in the Lake Chapala area collected in 2006–2007 (**Figure 7**). The river runoff data were obtained from CONAGUA (Mexico's National Water Council). The wet-season average of the inflow from the Lerma River was set to 600cubic meters per second, the outflow through the Santiago River —120 cubic meters per second.

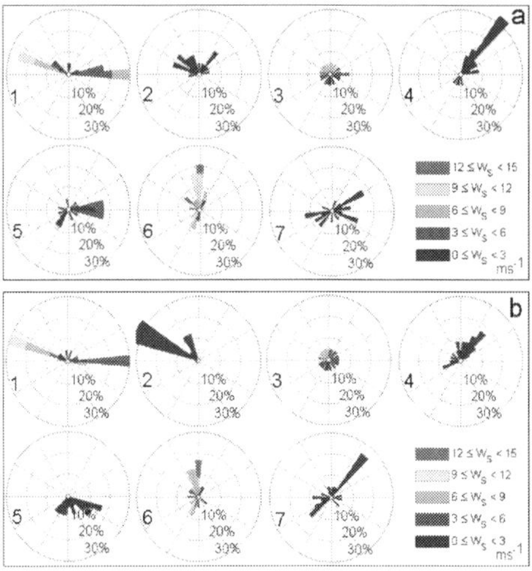

Figure 7. Annual wind rose at the weather stations around the lake: (a) the dry season; (b) the wet season. The data were averaged over the years 2006–2007.

The model results are shown in **Figure 8** as vector plots of current fields at 4-hour interval. The breeze-induced circulation pattern in the lake is represented by two gyres. One of them is cyclonic (counterclockwise rotation) and located in the east-central part of the lake. The other gyre is anticyclonic (clockwise rotation) and located in the west-central part. The model results exhibit a very complex dynamics, whose small-scale features are difficult to interpret. Nevertheless, similar gyres were identified near the east and west coast.

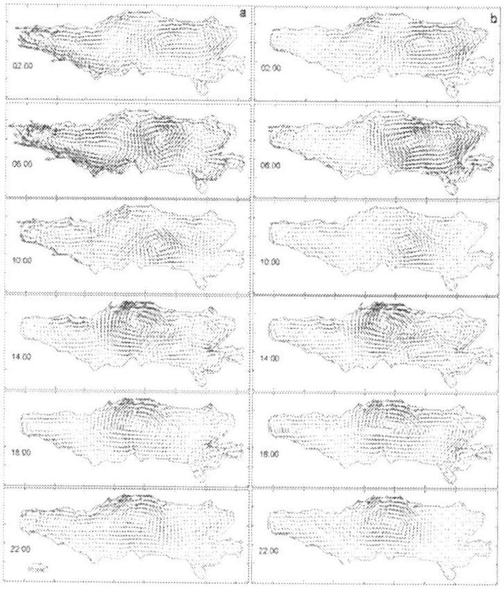

Figure 8. Model simulations of the wind-generated: (a) the dry season; (b) the wet season. The data were averaged over the years 2006–2007.

The core of the anticyclonic gyre, which was generated in the west part of the lake, propagates to the northwest and continues to develop during the most part of the simulation period. Conversely, the cyclonic gyre originally located in the east-central part moves to the southwest part of the lake. Subsequently, it vanishes due to the bottom friction and its remainder merges into the returning flow of the anticyclonic gyre. The model simulations show that the currents near the south and north coast reach 12 cm/s. These and other model results require the comparisons with current meter data. Therefore, two special experiments were carried out in January 2007 and June 2014.

3.6.1. *Temperature and currents variability within two lake polygons*

In order to quantitatively describe the spatial-temporal variability of temperature and circulation in the Lake Chapala, two special experiments were conducted: (i) on January 10, 2007 (the Alacranes polygon, **Figure 9a**, center-left) and (ii) on June 1, 2014 (the Mezcala polygon, **Figure 9a**, center). Both experiments were conducted with the use of a vertical array of temperature recorders and a boat-mounted ADCP.

The array contained 15 temperature recorders (13 HOBO Pro v2, one HOBO-LEVEL sensor and one SBE-39), evenly placed from the surface to 7.2 m depth. The boat towed the array. A down-looking ADCP RDI 600 kHz set up in bottom-track mode was mounted on the starboard side of the boat. The bin size was set to 15 cm (the total of 13 bins), and the ensemble interval was 15 s.

The instruments were protected from fishing nets by a specially designed triangular metal case shaped to fit the recorders so that the fishing gear caused no damage. The designated depth of the array was controlled by HOBO-LEVEL and SBE-39 pressure sensors and weights (**Figure 9b**). The sampling rate of temperature and pressure sensors was 1 min. The towing speed of about 2 m/s allowed the acquisition of temperature and currents data with the horizontal spatial resolutions of 120 and 30 m, respectively.

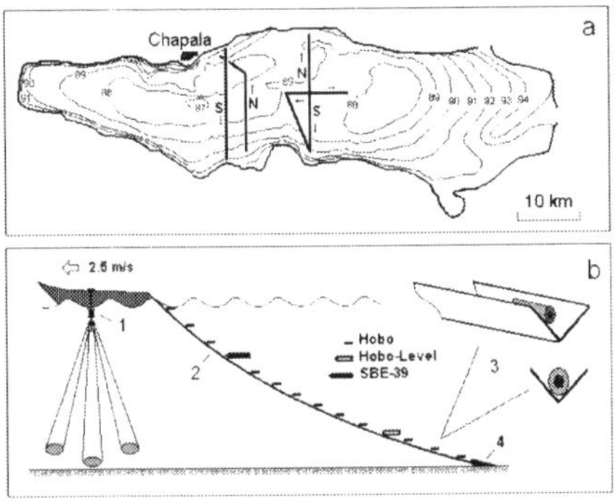

Figure 9. (a) The schematic of the experiments in the Lake Chapala conducted on January 10, 2007. (polygon Alacranes: center-left) and June 1, 2014 (polygon Mezcala: center of the figure). (b) The schematic of the towed array of recorders. 1. 600 kHz RDI ADCP, 2. angular metal case for the protection of the recorders, 3. the recorder mount and 4. weights for lowering of the recorders.

Current profiler data is calibrated according to [34], and bad data removed following the procedure explained in Ref. [35].

3.6.2. The Alacranes polygon

The spatial distribution of temperature along the transect **S** (carried out from 12:00 to 14:30, the boat sailed southward) is shown in **Figure 10a**. The transect *N* (on which the boat sailed from south to north) was carried out from 16:40 to 17:50 and the corresponding temperature field is shown in **Figure 10b**. On both transects, the meteorological conditions were recorded. During the transect **S**, the average air temperature was 14.5°C; the wind was blowing onshore, and its speed was 12 m/s. Later, during the transect **N**, the average air temperature was higher and reached 20.9°C; the wind was blowing offshore, which is a typical breeze circulation in the Lake Chapala [18].

It is seen from these transects that the heat fluxes were directed toward the surface layer of the water column, whereby the vertical gradient reached 2.5°C per the top meter of the column. During the cross section **S**, all temperature fluctuations were confined between 17 and 18.5°C. Few hours later, the rapid development of a thermocline was observed on transect *b* (**Figure 10**) with temperatures ranging between 17 and 20°C. At the end of the **N** transect, the coastal water began to mix due to the lake breeze effect. Furthermore, at the north end of the transectS the water was warmer than at the south by 1°C (**Figure 10**).

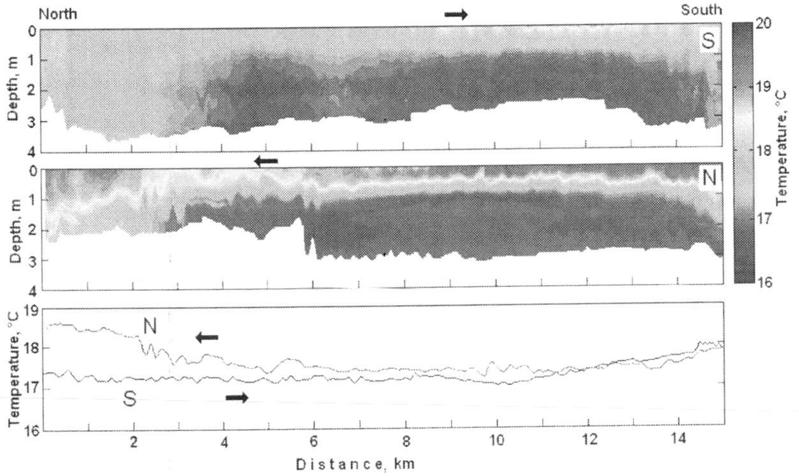

Figure 10. Vertical temperature cross sections of the lake carried out from north to south (S) and from south to north (N) obtained from the arrays of recorders on January 10, 2007, in the Lake Chapala. The bottom panel shows the vertically averaged temperatures at both cross sections.

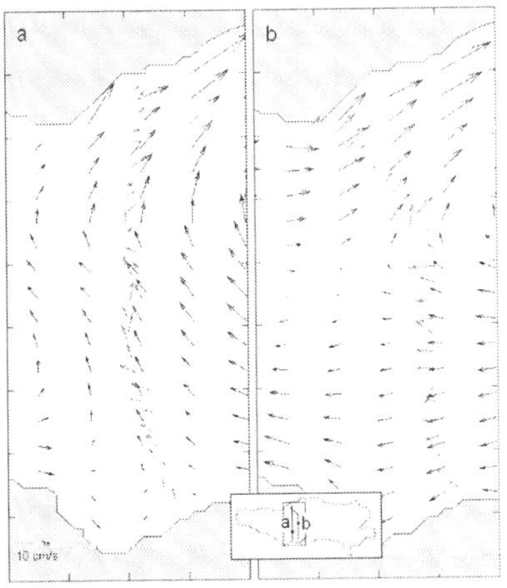

Figure 11. (a) Modeled (black) and observed (red) currents in the vicinities of the *S* and *N* transects. The experiment was conducted on January 10, 2007; the simulations are shown for 12:30. and 19:00, respectively.

The wind field is believed to be the main mechanism that continually sustains the circulation in the Lake Chapala, including the gyres. **Figure 11** presents the vector plots of vertically averaged currents measured by the towed ADCP and simulated by the HAMSOM model. In general, the modeled and observed data are in good agreement.

The transect ***S*** was carried out from 12:00 to 14:30, while the boat was sailing from north to south. The speed of the northeastward flow along the cross section peaked at 10 cm/s. The flow was presumably generated by the prevailing onshore wind dubbed "the Mexican" by local fishermen [22]. The 13:00 model simulations are in good agreement with the data collected in the northern part of the lake (**Figure 8a**). At the same time, the south part is characterized with a significant difference between the modeled and observed data, in terms of both the currents and temperature field.

Nevertheless, the transect ***N*** data collected between 16:30 and 18:00 show fairly good agreement with the southern part of the 18:00 model field, with velocities reaching 15 cm/s. It was found that in the southern part of the lake, the direction of the flow could change from west to south in only 2 hours. The wind speed of 12 m/s was recorded by the Chapala city weather station at the same time when transect ***S*** displayed southeastward flow. In just 2 hours upon the completion of transect *N*, the flow changed its direction to southeastward and gained speed of 10 cm/s. These results suggest that the model successfully simulates the effect of the morning breeze circulation.

3.6.3. *The Mezcala polygon*

The model simulations show the presence of a steady anticyclonic gyre of 10–12 km diameter in the central part of the lake, across from the town of Mezcala, during both seasons (**Figure 8**). To confirm these numerical calculations, we carried out the Mezcala polygon survey. The temperature and currents within the gyre where observed with the use of towed temperature recorders and ADCP. The cross-shaped polygon of about 6 km length was situated in the deepest part of the lake, inside the gyre. Continuous measurements along the three directions (**Figure 9a**) were taken on July 1, 2014 from 8:30 to 19:00, whereby the boat was sailing back and forth.

As shown in **Figure 12**, the temperature is distributed along the two transects: the transect *S*, which was carried out from 8:30 to 10:20, while the boat was sailing south and the transect *N*, which were carried out from 17:00 to 19:00. The meteorological conditions during the experiment were typical for this time of the year: the morning was calm followed by a moderate breeze in the afternoon. The towing speed was higher than on the Alacranes polygon; therefore, the deepest recorder was only at a 2.2 m depth. Nevertheless, the collected data shed a new light on the horizontal and vertical structure of the temperature field on the polygon in the presence of the anticyclonic gyre.

As seen from **Figure 12**, in the morning (cross section *S*), at the time when the boat was sailing south (from 9:15 to 10:22 AM), the stratification of the lake was moderate. The average temperature near the southern shore of the lake was steady at 23.4°C (lower panel at the **Figure 12**). During the reverse leg along the transect (from the south to the north), the measurements were taken only in the afternoon (from 17:00 to 19:00), when the wind and temperature conditions over the lake were quite different from the morning. The simulated currents are reasonably comparable with the observed data and suggest that in the second half of the day the gyre was well developed. The gyre exerts a strong impact on the spatial temperature distribution in the study area.

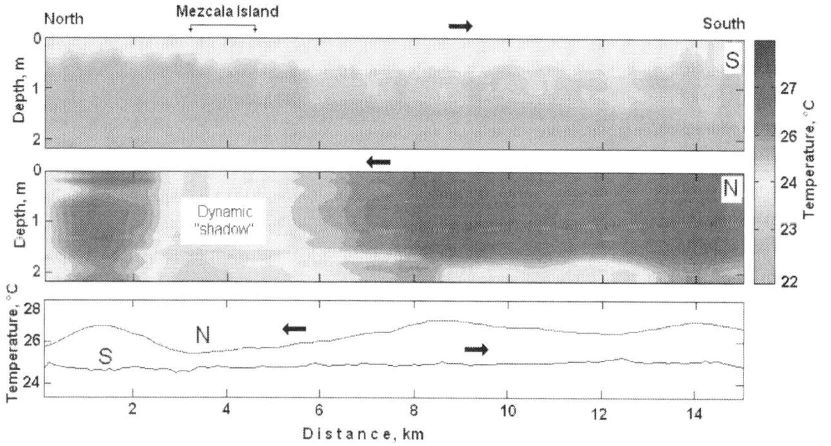

Figure 12. Vertical temperature cross sections: the north-to-south (*S*) and south-to-north (*N*) transects, obtained from the arrays of recorders on July 1, 2014. The vertically averaged temperature at both cross-sections is shown on the lower panel.

A zone of strong vertical mixing was discovered on the last leg, about 1 km east from the Mezcala Islands, when the array of the recorders was towed from the south to the north. These islands are about 1.5 km long. They are perpendicular to the flow of the gyre and create dynamic shadows. Within this area, the upwelling of cold-bottom water due to the Bernoulli effect resulted in its mixing with the warm surface water. Unfortunately, the spatial grid of the model (300 × 300 m) was too coarse to simulate this effect.

More detailed hydrographic surveys off the Mezcala Islands and the Alacranes polygon are planned for the future with the subsequent assimilation of the collected data into 3D numerical models. This will allow us to confirm the above-mentioned assumptions about the impact of the vortex on the vertical and horizontal mixing near the islands.

4. Conclusions

A great deal of results presented here are unique for the Lake Chapala. Although this study was carried out during different time periods, on average, it gives a fair account of the dynamic processes occurring in the lake and its surroundings. The main dynamic process occurring in the lake is the breeze circulation. The daytime breeze does not usually exceed 4 m/s. Beyond any doubt, the lake breeze results in the increased evaporation from the lake's surface and also generates free seiches waves.

The spectral analysis of the lake-level fluctuation measured by a high-precision HOBO-level recorder shows that there are two seiches modes in the lake with the periods of 5.7 and 2.8 hours, with the average amplitudes of 15.4 and 8.1 mm.

This study shows for the first time the main features of the diurnal variability of the stratification in the lake. The thermocline is formed around midday. It is very narrow and well defined. Within the top 30–40-cm layer, the vertical temperature gradient reaches 0.5°C/m and falls to zero at the bottom. The buoyancy frequency in the surface layer during the day hours exceeds 50–60 c/hour (a period of about 1 min). The well-defined stratification of the surface layer prevents the development of turbulence and mixing by wind and waves. In contrast, during the nighttime, negative temperature gradients of up to 3°C/m are formed due to the strong cooling of the surface layer. This serves to increase the intensive vertical mixing and results in a constant temperature within the entire water column of the lake.

The new data on the diurnal variations of the thermal structure of the lake are very important for the understanding of the migration behavior of biological organisms. The data are also important for the development of three-dimensional models of the dynamic processes in the lake.

The study shows that the northern part of the lake is warmer than its central and southern parts, which is essentially an important feature of the circulation. However, until now there were no detailed accounts of the horizontal and vertical structure of the temperature field in the lake in different seasons.

Our *in situ* measurements taken in February, April, July and October 2006 confirmed the previous finding that the temperature in the central and northern parts of the lake during different seasons is always higher than in the southern part, which holds true not only for the surface but also for the entire water column. In all four seasons, the average temperatures at the north and south coasts differ by 2–3°C. The vertical distribution of temperature along the cross sections was not homogeneous.

In this study, the simulation of currents was carried out by means of the HAMSOM 2D model for the dry and wet seasons. The model was initialized with the wind data from the network of weather stations in the Chapala Lake area collected in 2006–2007. The model results demonstrate very complex dynamics, in particular, the continuous presence of two gyres. One of them rotates counterclockwise (cyclonic rotation) and is located in the east-central part of the lake. The other one rotates clockwise (anticyclonic rotation) and is located in the west-central part.

In order to describe the spatial-temporal variability of temperature in the lake and compare the model simulations with the observed data, two special experiments were conducted in the Lake Chapala on January 10, 2007 (polygon Alacranes) and on June 1, 2014 (polygon Mezcala). In these experiments, a vertical array of temperature recorders aligned in the antenna-like pattern was towed along the cross sections by a boat with onboard ADCP. These high-frequency measurements on both polygons shed new light on the distribution of temperature and currents in these parts of the lake. The data collected on the Mezcala polygon confirm the presence of an anticyclonic gyre and show the influence of the islands on the dynamics of water masses and the temperature distribution in the lake.

Acknowledgements

This work was supported by the Mexican National Council for Science and Technology (CONACYT projects 33667-T and 41667).

Author details

Filonov Anatoliy[1*], Iryna Tereshchenko[1], Cesar Monzon[1], David Avalos-Cueva[2] and Diego Pantoja-González[1]

*Address all correspondence to: afilonov@prodigy.net.mx

1 Physics Department, University of Guadalajara, Marcelino Garcia Barragan Blvd., Guadalajara, Jalisco

2 Higher Technological Institute of Arandas. C.P., Arandas, Jalisco

References

[1] Sandoval, F. The past and future of Lake Chapala [in Spanish]. Government of the State of Jalisco, General Secretary, Editorial Unit, Jalisco, Mexico; 1994. 94 p.

[2] Tereshchenko IE, Filonov AE, Gallegos A, Monzón CO, Rodríguez R. El Niño 1997–98 and the hydrometeorological variability of Chapala, a shallow tropical lake in Mexico. Journal of Hydrology. 2002;264:133–146.

[3] Avalos-Cueva D, Filonov AE, Tereshchenko IE, Monzón CO, Pantoja-González D, Velázquez-Muñoz FA. The level variability, thermal structure and currents in Lake Chapala, Mexico. Journal Geofísica Internacional. 2016;56:XX–XX.

[4] Jáuregui E. Rainfall fluctuations and tropical storm activity in Mexico. Erdkunde. 1995;49:39–48. Available from: http://www.jstor.org/stable/25646634

[5] Filonov AE. Tereshchenko IE. Preliminary results on the thermic regime of Lake Chapala. Mexico. Supplement to EOS, Transact, AGU. 1997;78:46.

[6] Filonov AE, Tereshchenko IE, Monzón CO. Oscillations of the hydrometeorological characteristics in the region of Lake Chapala for intervals of days to decades. Geofísica Internacional. 1998;37:393–308.

[7] Riehl H. Climate and Weather of the Tropics. Academic Press, London; 1979. 611 p.

[8] Filonov AE. Will die the Lake Chapala? There is the possibility of being saved [in Spanish]. Teorema 1998;3:16–18.

[9] Filatov NN. Dynamics of lakes. Gidrometeoizdat, Leningrad; 1983. 161 p. [in Russian].

[10] Magaña, VO, Vázquez JL, Pérez JL, Pérez JB. Impact of El Niño on precipitation in Mexico. Geofísica Internacional. 2003;42(3):313–330.

[11] Soto-Mardones L, Marinone SG, Pares-Sierra A. Spatio-temporal variability of sea surface temperature in the Gulf of California [in Spanish]. Ciencias Marinas. 1999;25:1–30.

[12] Schwab DJ, Leshkevich GA, Muhr GC. Satellite measurements of surface water temperature in the Great Lakes: Great Lakes Coastwatch. Journal of Great Lakes Research, 1992;18(2):247–258.

[13] Strub PT, James C. Altimeter-derived variability of surface velocities in the California Current System: 2. Seasonal circulation and eddy statistics. Deep Sea Research Part II: Topical Studies in Oceanography. 2000;47(5):831–870.

[14] Avalos-Cueva D, Filonov AE, Tereshchenko IE, Monzón CO, Velázquez-Muñoz FA. Thermal structure and circulation in Lake Chapala, Mexico. Journal Limnology. 2016;75(s1):137–143. DOI: 10.4081/jlimnol.2015.1430.

[15] LeBlond PH, Mysak LA. Waves in the Ocean. Elsevier Oceanography Series 20, Amsterdam; 1978. 602 p.

[16] Graf, WH, Mortimer CH, editors. Hydrodynamics of Lakes. Elsevier Amsterdam-Oxford-New York; 1979;11. 359 p.

[17] Scorer RS. Environmental Aerodynamics. John Wiley, New York; 1978. 523 p.

[18] Filonov AE. On the dynamical response of Lake Chapala, Mexico to lake breeze forcing. Hidrobiología. 2002;467:141–157.

[19] De-Anda J. Solids distribution in Lake Chapala, México. Journal of the American Water Resources Association. 2004;40:97–109.

[20] Filonov AE, Tereshchenko IE, Alcocer J, Monzón CO. Dynamics of internal waves generated by mountain breeze in Alchichica Crater Lake, Mexico. Geofísica Internacional. 2015;54(1):21–30.

[21] Filonov AE, Tereshchenko IE. Thermal fronts and internal nonlinear waves in shallow tropical Lake Chapala, Mexico. Russian Meteorology and Hydrology. 1999:1:58–64.

[22] Avalos-Cueva D. Breeze circulation and influence the thermal structure of Lake Chapala, Mexico [master's thesis in Spanish]. Master in Earth sciences; Department of Physics: University of Guadalajara; 2003.

[23] Filonov AE, Tereshchenko IE. Thermal lenses and internal solitones in the shallow lake Chapala, Mexico. Chinese Journal of Oceanology and Limnology. 1999;17:308–314.

[24] Filonov AE, Tereshchenko IE, Monzón CO. Hydrometeorology of Lake Chapala. In: Hansen AM, van Afferden M, editors. The Lerma-Chapala Watershed: Evaluation and Management. Kluwer Academic Publishers/Plenum Press, New York; 2001. pp. 151–182.

[25] Simons TJ. Effect of outflow diversion on circulation and water quality of Lake Chapala. Report Project MKX CWS-01. 1984. p. 23.

[26] Escalante ME. Pollutant transport in Lake Chapala (initial stage) [in Spanish]. Technical report CH-9206/07. 1992.

[27] Arakawa A, Lamb VR. Computational design of the basic dynamical processes of the UCLA general circulation model. Methods in Computational Physics. 1977;17:173–265.

[28] Carbajal N. Modeling of the Circulation in the Gulf of California [thesis]. Institute of Oceanography: University Hamburg; 1993.

[29] Marinone SG. A three-dimensional model of the mean and seasonal circulation of the Gulf of California. Journal of Geophysical Research: Oceans. 2003;108:C103325.

[30] Serrano D, Filonov AE, Tereshchenko IE. Dynamic response to valley breeze circulation in Santa Maria del Oro, a volcanic lake in Mexico. Geophysical Research Letters. 2002;29:27–31.

[31] Filonov AE, Tereshchenko IE, Alcocer J. Dynamic response to mountain breeze circulation in Alchichica, a crater lake in Mexico. Geophysical Research Letters. 2006;33:L07404. DOI: 10.1029/2006GL025901.

[32] Hainbucher D, Pohlmann T, Backhaus J. Transport of conservative passive tracers in the North Sea: first results of a circulation and transport model. Continental Shelf Research. 1987;7(10):1161–1179.

[33] Ilyina T, Pohlmann T, Lammel G, Sündermann J. A fate and transport ocean model for persistent organic pollutants and its application to the North Sea. Journal of Marine Systems. 2006;63(1):1–19.

[34] Trump CL, Marmorino GO. Calibrating a gyrocompass using ADCP and DGPS data. Journal of Atmospheric and Oceanic Technology. 1997;14:211–214.

[35] Valle-Levinson A, Atkinson L. Spatial gradients in the flow over an estuarine channel. Estuaries and Costs. 1999;22:179–193.

Chapter 2

Relationship between Water Levels in the North American Great Lakes and Climate Indices

Ali A. Assani, Ouassila Azouaoui,
Anthony Pothier-Champagne and
Jean-François Quessy

Additional information is available at the end of the chapter

http://dx.doi.org/10.5772/63383

Abstract

The goal of this study is to look at how the interconnection of five North American Great Lakes affects the relationship between climate indices and mean annual and extreme daily water levels during the period from 1918 to 2012, and how human activity impacts the dependence between these two variables. Analysis of correlation revealed the existence of a negative correlation between water levels in Lakes Superior, Michigan–Huron and Erie, and the Atlantic Multidecadal Oscillation (AMO) climate index, although this correlation is not observed at the daily scale for Lake Superior. Water levels in Lake Ontario are negatively correlated with Pacific Decadal Oscillation (PDO). The temporal evolution of the dependence between water levels and climate indices is characterized by breaks interpreted to result from variations in the amount of precipitation probably linked with an AMO phase change in the Lakes Superior, Michigan–Huron, and Erie watersheds. In the case of Lake Ontario, such breaks in dependence are thought to be related to water level regulation in this lake resulting from the digging of the St. Lawrence Seaway.

Keywords: water levels, climate indices, correlation, copula, Great Lakes

1. Introduction

The North American Great Lakes system is one of the largest bodies of freshwater in the world. The system holds nearly 23,000 km^3 of water or about 20% of the World's freshwater reserves [1]. It is a rich and diverse aquatic ecosystem and continues to play a crucial role in the social and economic development of interior regions of the United States and Canada. For these reasons, the Great Lakes are the subject of numerous multidisciplinary studies. From a hydroclimate standpoint, most of these studies have focused primarily on the following four elements:

- the variability of hydroclimate variables (water levels, temperature, precipitation, evaporation, etc.) at different time scales (hourly, daily, seasonal, annual and secular) as it relates to natural and human factors [2–16, 17].
- the potential impacts of climate change on this variability, a topic of growing attention given the current climate warming [17–24, 25].
- the seasonal and interannual ice dynamics and its impacts on temperature and water levels in the Great Lakes [26–29, 30].
- the interaction of this water body with regional and/or global climate [31–33].

Few studies have attempted to determine which climate indices affect the interannual variability of water levels [2, 3, 12, 26, 29, 34], and most of these studies only focused on the relationship between climate indices and extreme water levels. To fill this gap, the first goal of the study is to analyze the relationship between climate indices and annual mean and daily extreme (maximum and minimum) water levels in five North American Great Lakes. Underlying this goal is the following hypothesis: because of their interconnected nature, water levels in the Great Lakes are correlated with each other and, as a result, are correlated with the same climate indices. In addition, given that some of the Great Lakes are affected to varying degrees by water level regulation [7, 35, 36], it might be expected that this regulation affects the dependence between water levels and climate indices over time. The second goal of this paper is therefore to analyze this change in dependence over time.

2. Methodology

2.1. Location and sources of data

Located almost entirely along the Canada–US border, the North American Great Lakes system comprises five large lakes (Superior, Michigan, Huron, Erie and Ontario) and a plethora of smaller ones (**Figure 1**). These large lakes vary in surface area from 82,367 km^2 (Superior) to 19,009 km^2 (Ontario), and in volume from 12,221 km^3 (Superior) to 458 km^3 (Erie Water residence time ranges from 191 (Superior) to 2.6 years (Erie).

Water level data were taken from the Environment Canada web site (http://www.waterlevels.gc.ca/C&A/network_means.html). It is important to note that from a hydraulic standpoint, Lakes Michigan and Huron form a single system (Michigan–Huron), and their water levels fluctuate in identical fashion. As a result, these fluctuations are measured at a single station, on Lake Michigan. For each lake, water level was correlated with the five climate indices which have been shown to affect climate in North America. These are: Atlantic Multidecadal Oscillation (AMO), Arctic Oscillation (AO), North Atlantic Oscillation (NAO),

http://dx.doi.org/10.5772/63383

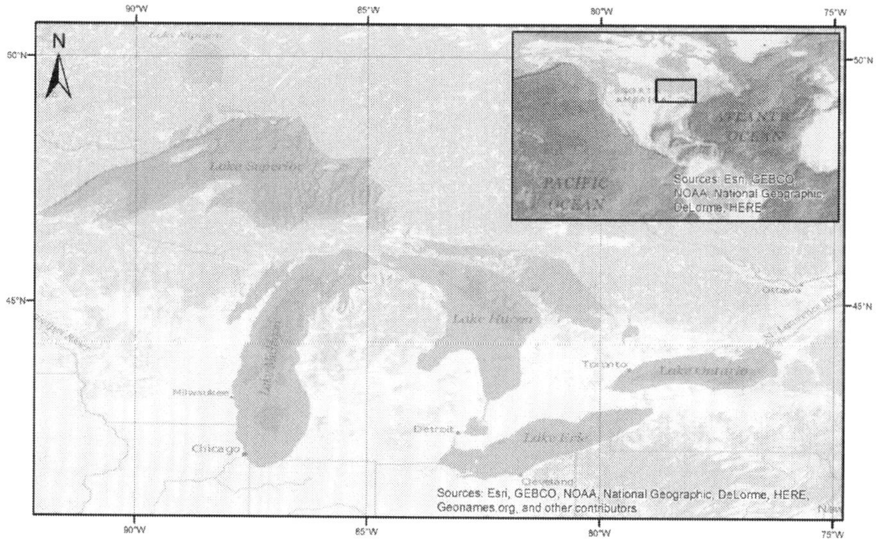

Figure 1. Location of the North American Great Lakes.

Pacific Decadal Oscillation (PDO), and Southern Oscillation Index (SOI). Climate indexes for the AMO and PDO were taken from the following website: http://www.cdc.noaa.gov/ClimateIndices/List, the NAO from http://www.cgd.ucar.edu/cas/jhurrell/indices.data.html, the AO from http://jisao.washington.edu/data/ao/, and the SOI from http://www.cgd.ucar.edu./cas/catalog/climind/soi.html. Data for these climate indices (NAO, AO and SOI) after 2006 were taken from the NOAA website: http://www.esrl.noaa.gov/psd/data/climateindices/list/. Data for all these indices since 1950 are available on the NOAA website: http://www.esrl.noaa.gov/psd/data/climateindices/list/.

For each of the five Great Lakes, the following series will be produced:

- A series of annual mean water levels (average of the 12 monthly values) for the period from 1918 to 2012.

- A series of annual daily maximum and minimum water levels consisting of the highest (maximum) and lowest (minimum) water level values measured each year from 1918 to 2012.

These series will be correlated with time series for five climate indices, including AMO, AO, NAO, PDO, and SOI. For each of these indices, the following three series will be produced:

- A series of annual means (average of the 12 monthly values) of the climate indices.

- A series of seasonal means in winter (average of monthly values of the 6 months from October to March) and in summer (average of monthly values of the 6 months from April to September).

2.2. Statistical analysis

2.2.1. *Analysis of correlation*

For statistical data analysis, we used the following programs: SAS Version 9.2 32 bits (correlation analysis and canonical correlation analysis), and Matlab Version R2013a (Lombard analysis). The statistical analysis was carried out in three steps. The first step consisted in deriving simple coefficients of correlation between water levels in five the Great Lakes in order to constrain the effect of their interconnected nature on the temporal variability of their water levels. As a second step, coefficients of correlation were derived between climate indices and water levels. These two approaches, however, cannot detect potential interactions between the five climate indices and the temporal variability of water levels. For this reason, canonical correlation analysis (CCA) [36] was used as a third step. This method allows the simultaneous analysis of correlation between two groups of variables.

If one wishes to calculate the relation between two groups, one of X variables (X1, X2,…, Xp) and the other of Y variables (Y1, Y2,…, Yq), one must calculate the canonical variables V (V1, V2,…,Vp) and (W1, W2,…, Wq), which are linear combinations of the X variables of the first group (in this case, the climatic indices) and the Y variables of the second group (in this case, the water levels of the five Great Lakes). Then, the canonical variables V and W are correlated between themselves, that is to say, V1 is correlated with W1, V2 to W2 and so on, in order to obtain the canonical correlation coefficients. After that, the canonical variables V and W are correlated with the variables X and Y, so as to obtain what are called structure coefficients. In fact, these coefficients measure the link (the correlation) between the canonical variables (V and W) and the original variables of groups X and Y. Thus, if X1 and X2 are correlated, for example, to V1, Y1 is correlated with W1. Y1 is therefore correlated with the original variables X1 and X2, since the canonical variables V1 and W1 are correlated. The main purpose of CCA is to maximize the correlations between the two groups of variables.

2.2.2. *The copula method*

To test the second hypothesis that underlies the study, the copula method will be applied to the series of water levels in the Great Lakes and the climate indices with which they are significantly correlated. This method is used to analyze the evolution over time of the dependence between two correlated variables by detecting significant breaks in Kendall's tau values. The timing of these breaks will be compared with the timing of construction of man-made structures carried out over time to regulate water levels in the Great Lakes. We described this method in some of our previous work [i.e., For *example*, see 2].

The dependence in a random vector (X, Y) is contained in its corresponding copula function C. Specifically, the celebrated theorem of Sklar ensures that there exists a unique $C:[0, 1]^2 \rightarrow [0, 1]$ such that

$$P(X \leq x, Y \leq y) = C\{P(X \leq x), P(Y \leq y)\}. \tag{1}$$

Quessy et al. [37] developed a testing procedure to identify a change in the copula (*i.e.*, dependence structure) of a bivariate series $(X_1, Y_1), \ldots, (X_n, Y_n)$. The idea is based on Kendall's tau, which is a nonparametric measure of dependence. Let $\hat{T}_{1:T}$ be Kendall's tau measured for the first T observations and $\hat{T}_{T+1:n}$ be Kendall's tau for the remaining $n - T$ observations. The proposed test statistic is

$$M_n = \max_{1<T<n} \frac{T(n-t)}{n\sqrt{n}} \left| \hat{T}_{1:T} - \hat{T}_{T+1:n} \right| \tag{2}$$

that is, a maximum weighted difference between the Kendall's tau. Since M_n depends on the unknown distribution of the observations, the so-called multiplier re-sampling method is used for the computation of p-values. Specifically, for n sufficiently large (n > 50), this method yields independent copies $M_n^{(1)}, \ldots M_n^{(N)}$ of M_n. Then, a valid p-value for the test is given by the proportion of $M_n^{(i)}$'s larger than M_n. For more details [i.e., F*or example,* see 37]. Usually, one can expect that the series $X_1, \ldots, X_n$ and $Y_1, \ldots, Y_n$ are subject to changes in the mean and/or variance following, for example, the smooth-change model [38]. If such changes are detected, the series must be stabilized (to remove the shift of the mean and variance) in order to have (approximately) constant means and variances. Finally, a change in the degree of dependence between two series is statistically significant when $M_n > V_c$, where V_c is the critical value derived from observational data. As part of this study, the copula method was applied to standardized water level and climate index data after removing any break in mean values in the hydroclimate series.

3. Results

3.1. Simple linear correlation analysis

The values of coefficients of correlation derived between water levels in each of the five Great Lakes are shown in **Table 1**, in which it may be seen that those values increase with decreasing distance between lakes. Thus, the lowest coefficient of correlation values are between Lakes Superior and Ontario. However, coefficients of correlation between Lakes Superior and Michigan–Huron, which are adjacent to one another, are lower than those observed between the other three lakes, which are closer to one another.

	Lake Superior	Lake Michigan	Lake Erie	Lake Ontario
Mean annual water level				
Lake Superior	1	0.6231	0.4059	0.3341
Lake Michigan		1	0.8613	0.7164
Lake Erie			1	0.8193
Lake Ontario				1
Annual maximum water level				
Lake Superior	1	0.5673	0.3783	0.3033
Lake Michigan		1	0.8442	0.7055
Lake Erie			1	0.8332
Lake Ontario				1
Annual minimum water level				
Lake Superior	1	0.6081	0.4110	0.2581
Lake Michigan		1	0.8261	0.6621
Lake Erie			1	0.7363
Lake Ontario				1

All coefficients of correlation are statistically significant at the 5% level.

Table 1. Correlation between water levels in the North American Great Lakes.

	Annual climatic indices				Winter climatic indices				Summer climatic indices			
	S	M–H	E	O	S	M–H	E	O	S	M–H	E	O
AMO	**−0.22**	**−0.42**	**−0.30**	−0.12	−0.19	**−0.42**	**−0.33**	−0.09	**−0.22**	**−0.38**	**−0.31**	−0.13
AO	−0.00	0.07	0.17	0.14	−0.01	0.07	0.15	0.12	0.01	0.02	0.08	0.10
NAO	0.07	0.05	−0.06	−0.04	0.07	0.16	0.18	0.08	0.02	−0.12	**−0.32**	−0.18
PDO	0.05	0.03	−0.01	**−0.22**	0.02	0.00	−0.06	**−0.28**	0.07	0.05	0.05	−0.16
SOI	−0.05	−0.09	−0.08	0.02	−0.06	−0.10	−0.07	0.05	−0.01	−0.04	−0.05	−0.02

Significant coefficient of correlation values at the 5% level are shown in bold.

Table 2. Coefficients of correlation calculated between the five climate indices and annual mean water levels (1918–2012).

As far as the relationship between climate indices and annual mean water levels is concerned, **Table 2** and **Figure 2** shows that AMO is negatively correlated with water levels in Lakes Superior, Michigan–Huron, and Erie, with water levels in Lake Michigan–Huron showing better correlation with this climate index than water levels in the other two lakes. Annual mean water levels in Lake Ontario are negatively correlated with PDO (**Figure 3**), the same type of relationship being observed for both maximum (**Table 3**) and minimum (**Table 4**) daily water levels. Aside from PDO, daily extreme water levels in Lake Erie are also negatively correlated with the NAO summer indices. However, daily extreme (maximum and minimum) water levels in Lake Superior are not significantly correlated with any climate index.

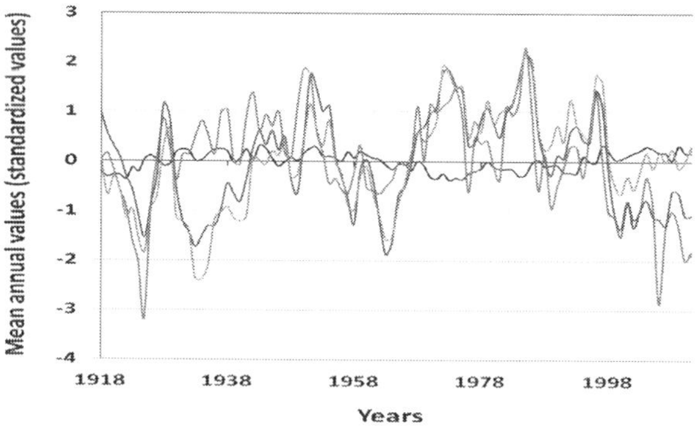

Figure 2. Comparison of the temporal variability of AMO mean annual indices (black curve) and mean annual water levels (standardized values) in Lakes Superior (red curve), Michigan–Huron (blue curve) and Erie (green curve) (1918–2012).

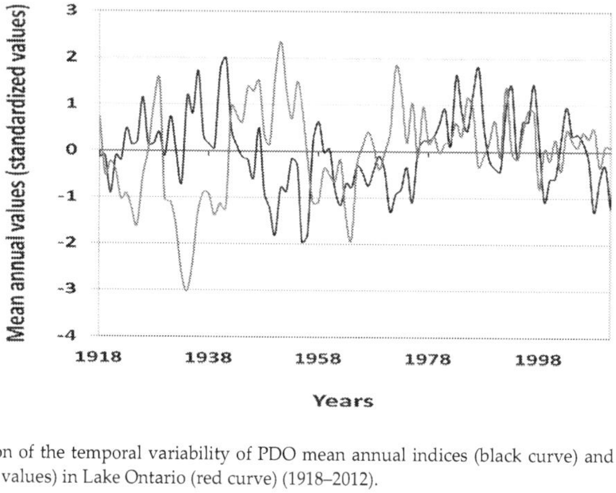

Figure 3. Comparison of the temporal variability of PDO mean annual indices (black curve) and mean annual water levels (standardized values) in Lake Ontario (red curve) (1918–2012).

	Annual climatic indices				Winter climatic indices				Summer climatic indices			
	S	M–H	E	O	S	M–H	E	O	S	M–H	E	O
AMO	−0.14	**−0.39**	**−0.23**	−0.10	−0.15	**−0.40**	**−0.27**	−0.12	0.04	0.08	0.10	0.14
AO	−0.09	0.09	0.12	0.08	−0.12	0.07	0.11	0.04	0.04	0.08	0.10	0.14
NAO	0.05	0.10	−0.03	0.01	0.04	0.19	0.20	0.13	0.03	−0.09	**−0.29**	−0.14
PDO	0.03	−0.01	−0.07	**−0.26**	0.02	−0.03	−0.11	**−0.29**	0.03	0.00	−0.02	**−0.21**
SOI	−0.15	−0.05	−0.02	−0.00	−0.11	**−0.34**	**−0.24**	−0.08	−0.10	−0.01	0.02	0.01

Significant coefficient of correlation values at the 5% level is shown in bold.

Table 3. Coefficients of correlation calculated between the five climate indices and maximum daily water levels (1918–2012).

	Annual climatic indices				Winter climatic indices				Summer climatic indices			
	S	M–H	E	O	S	M–H	E	O	S	M–H	E	O
AMO	−0.11	**−0.41**	**−0.31**	−0.12	−0.19	**−0.45**	**−0.35**	−0.10	−0.11	**−0.41**	**−0.31**	−0.12
AO	0.11	0.07	0.07	0.08	0.01	0.10	0.12	0.07	0.11	0.07	0.07	0.08
NAO	0.09	−0.06	−0.31	−0.20	0.09	0.20	0.14	0.08	0.09	−0.06	**−0.31**	−0.20
PDO	0.13	0.07	0.01	**−0.22**	0.12	0.04	−0.07	**−0.28**	0.13	0.07	0.01	**−0.22**
SOI	−0.07	−0.02	−0.01	−0.03	−0.12	−0.13	−0.11	−0.07	−0.07	−0.02	−0.01	−0.03

Significant coefficient of correlation values at the 5% level is shown in bold.

Table 4. Coefficients of correlation calculated between the five climate indices and minimum daily water levels (1918–2012).

3.2. Canonical analysis of correlation (CCA)

This analysis did not yield conclusive result. As an example, CCA results applied to annual mean water levels and annual mean climate indices are presented in **Tables 5** and **6**. From **Table 5**, it can be seen that the coefficient of canonical correlation between the first two axes is relatively low (0.633). CCA could not significantly maximize the coefficient of correlation values. Thus, the derived value of 0.633 is only slightly higher that the highest simple correlation coefficient of 0.425 (see **Table 2**). In addition, only the first two canonical axes are statistically significant at the 5% level. As far as coefficients of structure are concerned (**Table 6**), values of coefficients of structure related to water levels in the lakes are all lower than 0.600 on the first canonical axis. Thus, this axis shows little correlation with water levels in the lakes. The same is true for climate indices on the second canonical axis. As the last two canonical axes are not statistically significant, it is difficult to interpret their coefficients of structure.

Canonical roots	R	F	p-values
CC1	0.6330	3.81	<0.0001
CC2	0.4211	2.11	0.0171
CC3	0.2832	1.24	0.2872
CC4	0.0156	0.9997	0.9894

R = canonical coefficient of correlation.

Table 5. Canonical correlation analysis statistics.

Variables	W1	W2	W3	W4	V1	V2	V3	V4
Lake Superior	−0.370	0.003	0.258	0.893				
Lake Michigan–Huron	−0.587	0.263	0.725	0.245				
Lake Erie	−0.293	0.682	0.655	0.142				
Lake Ontario	0.133	0.346	0.914	0.164				
AMO					0.750	−0.195	−0.606	−0.033
AO					0.070	0.471	0.238	0.837
NAO					−0.195	−0.418	0.105	0.879
PDO					−0.470	0.200	−0.829	0.204
SOI					0.257	−0.157	0.093	−0.058

Table 6. Correlation between the annual mean water levels and canonical roots (W), and correlation between climatic indices and canonical roots (V).

3.3. Analysis of the dependence between water levels and climate indices using copulas

It is important to point out that the copula method was only applied for climate indices that are significantly correlated with water levels. Results obtained using this method are presented in **Table 7**, from which it may be seen that the dependence between the two climate indices (AMO and PDO) and water levels in the five Great Lakes, aside from annual mean water levels in Lake Ontario, shows a sharp break (**Figures 4–6**). In other words, the relationship between the two variables changed significantly over time.

As far as the timing of this change in dependence is concerned, **Table 7** shows that it is nearly synchronous for Lakes Michigan–Huron and Erie, having occurred in the late 1960s and early 1970s. In Lake Ontario, these breaks occurred in the late 1950s. Finally, for Lake Superior, the break in mean water levels took place during the first half of the 1950s. In all cases, the value of Kendall's tau decreases after the break.

Lakes	Indices	M_n	V_c	p-value	Year
Mean annual water levels					
Lake Superior	AMO	1.0614	0.9389	0.020	1954
Lake Michigan	AMO	1.2809	0.8337	0.001	1966
Lake Erie	AMO	1.3728	0.8204	0.000	1968
Lake Ontario	PDO	0.857	0.904	0.075	1960
Annual daily maximum water levels					
Lake Superior	–	–	–	–	–
Lake Michigan	AMO	0.9715	0.8972	0.0270	1968
Lake Erie	AMO	1.1435	0.9293	0.010	1972
Lake Ontario	PDO	0.994	0.8865	0.023	1959
Annual daily minimum water levels					
Lake Superior	–	–	–	–	–
Lake Michigan	AMO	1.1311	0.8636	0.000	1968
Lake Erie	AMO	1.3756	0.9205	0.000	1968
Lake Ontario	PDO	1.1311	0.9040	0.000	1959

p-values < 0.05 are statistically significant at the 5% level.
* Year of break in dependence.

Table 7. Analysis of the relationship between climate indices and water levels using copulas.

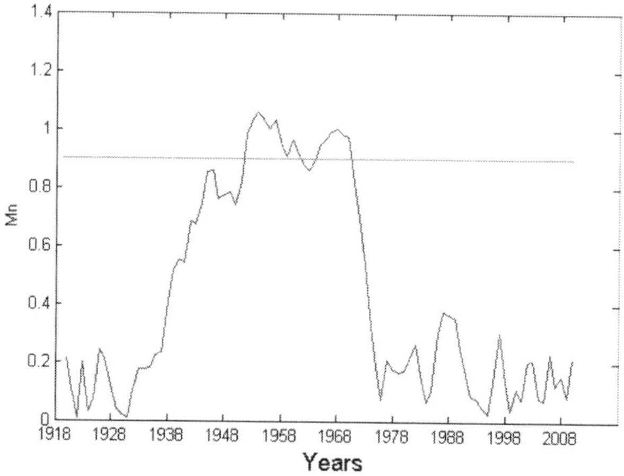

Figure 4. Interannual variability of M_n values of mean water levels. The red line represents the maximum value of V_c. Lake Superior.

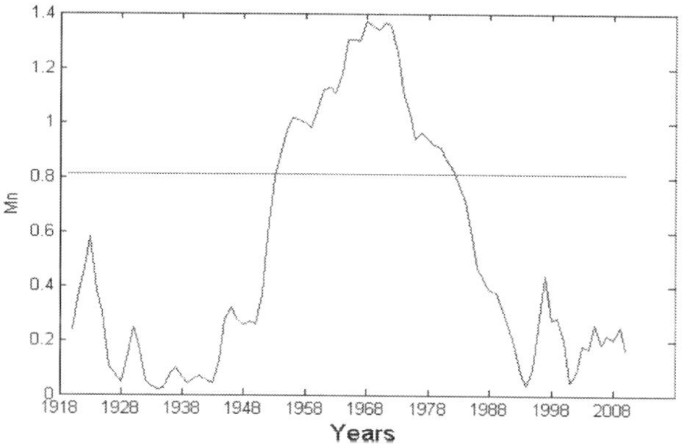

Figure 5. Interannual variability of M_n values of mean water levels. The red line represents the maximum value of V_c. Lakes Michigan–Huron.

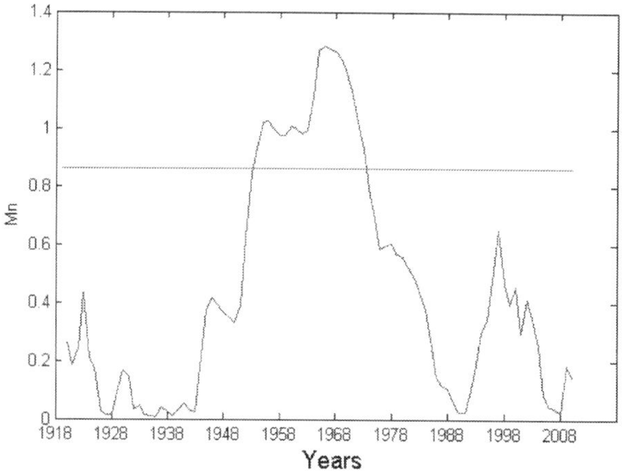

Figure 6. Interannual variability of M_n values of mean water levels. The red line represents the maximum value of V_c. Lake Erie.

4. Discussion and conclusion

The five North American Great Lakes comprise the largest freshwater aquatic ecosystem in the world. One of the characteristic features of these water bodies is their interconnected nature. Simple correlation analysis revealed that the effect of this interconnection on the temporal variability of water levels is strongly influenced by the distance between any two lakes. As this distance increases, the influence of the upstream lake on the temporal variability of water levels in the downstream lake decreases significantly. Thus, the weakest correlation between water levels is derived between Lakes Superior and Ontario; the two lakes located the furthest apart. In contrast, the strongest correlations are observed between adjacent lakes. This diminishing influence of interconnection with distance between lakes may in part be due to the dominant influence of local climate conditions in the watershed of each of the Great Lakes on the temporal variability of water levels [3].

The interconnected nature of the five lakes leads to the postulate that the interannual variability of their water levels may be correlated with the same climate indices. Analysis of the correlation between these water levels and five climate indices revealed a negative correlation between AMO and water levels in Lakes Superior, Michigan–Huron, and Erie, on one hand, and a negative correlation between PDO and water levels in Lake Ontario, on the other hand. It is worth specifying, however, that for Lake Superior, this correlation is only observed at the annual scale. While this result cannot be used to conclude that the interconnected nature of the lakes influences the relationship between climate indices and water levels, it does suggest a possible influence of the effects of water level regulation on this relationship. The two least regulated of the five lakes (Michigan–Huron and Erie) are correlated with the same climate indices at all time scales. Although Lake Superior is adjacent to Lake Michigan–Huron, its water levels are not correlated with AMO at all scale. And as far as Lake Ontario is concerned, despite its proximity to Lake Erie, its water levels are correlated to a different climate index, namely PDO.

The influence of AMO on the temporal variability of hydroclimate variables has been highlighted for many regions of North America [39–43, 44]. The sign of this correlation changes from region to region. For instance, in the Pacific Northwest and Northeastern regions of the United States, as well as in Florida, this correlation is positive, whereas in the continental region and the Mississippi River and Great Lakes watersheds, this correlation is negative [41]. Thus, during a positive phase of AMO, water levels in the Great Lakes tend to increase due to an increase in precipitation likely resulting from enhanced cyclonic activity in winter. In contrast, the influence of PDO on the temporal variability of hydroclimate variables has mainly been detected in the western part of the North American continent [45–49, 50]. In the Great Lakes watershed, a negative correlation between PDO and amount of snow has been brought to light [51, 52, 53], with the negative phase of this index being associated with an increase in amount of snow and, as a result, in water levels in Lake Ontario.

Over the years, the Great Lakes were affected to varying degree by human activity, including diversion within watersheds; dredging of natural channels connecting the lakes; and regulation of water levels [8, 54, 55]. It is therefore reasonable to expect that these different human impacts may have affected the dependence between climate indices and water levels in the lakes. Application of the copula method revealed breaks in this dependence at different scales. For water levels in Lakes Michigan–Huron and Erie, the three waterbodies least affected by human activity, breaks in this dependence occurred in the late 1960s and early 1970s. The only plausible factor that could account for these breaks is changing climate conditions in the watersheds of these three Great Lakes. Between 1970 and 1990, precipitations increased significantly in the Great Lakes watershed [e.g., *For example*, see 2], coinciding with a change in phase for AMO, from positive to negative after 1970 [41]. As far as Lake Ontario is concerned, the breaks in dependence occurred in the late 1950s and may therefore be related to water level regulation in this lake as a result of the digging of the Great Lakes-St. Lawrence Seaway System. Finally, for Lake Superior, the break in dependence occurred in 1954, which is not related to any significant development project. However, from a hydrological and climate standpoint, Assani et al. [2] observed an increase in the frequency of relatively low water levels during the 1950s and 1960s, reflecting a trend of decreasing precipitation in the watershed.

In conclusion, the influence of the interconnected nature of the Great Lakes on the temporal variability of their water levels decreases with increasing distance between individual lakes, likely as a result of the effect of local climate conditions on this variability. At the annual scale, however, water levels in four of the five Great Lakes are negatively correlated with AMO, and only water levels in Lake Ontario, the most downstream of the lakes, are correlated with PDO. For the first four lakes, the break in dependence between water levels and climate indices was affected by changes in climate conditions in their watersheds, whereas for Lake Ontario, water level regulation in its watershed affected this break. This study shows that the influence of local climate conditions and human activity in the watersheds of the Great Lakes strongly dampen the effects of interconnection on the temporal variability of water levels in the lakes. As a result, hydroclimate changes affecting one lake do not necessarily propagate to other lakes located downstream.

Author details

Ali A. Assani[1*], Ouassila Azouaoui[1], Anthony Pothier-Champagne[1] and Jean-François Quessy[2]

*Address all correspondence to: Ali.Assani@uqtr.ca

1 Department of Environmental Sciences, University of Quebec at Trois-Rivières, Boulevard des Forges, Trois-Rivières, QC, Canada

2 Department of Mathematical and Computer Sciences, University of Quebec at Trois-Rivières, Boulevard des Forges, Trois-Rivières, QC, Canada

References

[1] Magnuson JJ, Webster KE, Assel RA, Bowser CJ, Dillon PJ, Eaton JG, Evans HE, Fee EJ, Hall RI, Mortsh LR, Schindler DW, Quinn FH: Potential effects of climate changes on aquatic systems Laurentian Great Lakes and Precambrian Shield Region. Hydrological Processes. 1997; 11: 825–871.

[2] Assani AA, Landry R, Stacey B, Frenette JJ: Analysis of the interannual variability of annual daily extreme water levels in the St. Lawrence River and Lake Ontario from 1918 to 2010. Hydrological Processes. 2014; 28: 4011–4022.

[3] Assani AA, Landry R, Azouaoui O, Massicotte P, Gratton D: Comparison of the characteristics (frequency and timing) of drought and wetness indices of annual mean water levels in the five North American Great Lakes. Water Resources Management. 2016; 30: 359–373.

[4] Assel RA, Quinn FH, Sellinger CE: Hydroclimatic factors of the recent record drop in Laurentian great lakes water levels. Bulletin of American Meteorology Society.2004; 85: 1143–1151.

[5] Austin J, Colman S: A century of temperature variability in Lake Superior. Limnology and Oceanography. 2008; 53: 2724–2730.

[6] Cengiz TM: Periodic structures of great lakes levels using wavelet analysis. Journal of Hydrology and Hydromechanic. 2011; 59: 24–35.

[7] Changnon SA: Temporal behavior of levels of the Great Lakes and climate variability. Journal of Great Lakes Research. 2004; 30: 184–200.

[8] Clites AH, Quinn FH: The history of Lake Superior regulation: implications for the future. Journal of Great Lakes Research. 2003; 29: 157–171.

[9] Ehsanzandeh E, Saley HM, Ouarda TBMJ, Burn DH, Pietroniro A, Seidou O, Charron C, Lee D: Analysis in the Great Lakes hydro-climatic variables. Journal of Great Lakes Research. 2013; 39: 383–394.

[10] Hanrahan J, Roebber P, Kravtsov S: Attribution of decadal-scale lake-level trends in the Michigan–Huron system. Water. 2014; 6: 2278–2299.

[11] McBean E, Motiee H: Assessment of impact of climate change on water resources: a long term analysis of the Great Lakes of North America. Hydrology and Earth System Sciences. 2008; 12: 239–255.

[12] Ghanbari RZ, Bravo HR: Coherence between atmospheric teleconnections, Great Lakes water levels and regional climate. Advances in Water Resources. 2008; 31: 1284–1298.

[13] Gronewold AA, Fortin V, Logfren B, Clites A, Stow CA, Quinn F: Coasts, water levels, and climate change: a Great Lakes perspective. Climate Change. 2013; 120: 697–711.

[14] Lenters JD: Trends in the Lake Superior water budget since 1948: a weakening seasonal cycle. Journal of Great Lakes Research. 2004; 30: 20–40.

[15] Quinn FH: Secular changes in Great Lakes water level seasonal cycles. Journal Great Lakes Research. 2002; 28: 451–465.

[16] Quinn FH, Sellinger CE: A reconstruction of Lake Michigan–Huron water levels derived from tree ring chronologies for the period 1600–1961. Journal Great Lakes Research. 2006; 32: 29–39.

[17] Sellinger CE, Stow CA, Lamon EC, Qian SS: Recent water level declines in the Lake Michigan–Huron systems. Environmental Sciences and Technology. 2008; 42: 367–373.

[18] Angel JR, Kunkel KE: The response of great lakes water levels to future climate scenarios with an emphasis on Lake Michigan–Huron. Journal of Great Lakes Research. 2010; 36: 51–58.

[19] Cherkauer KA, Sinha T. Hydrologic impacts of projected future climate change in the Lake Michigan region. Journal Great Lakes Research. 2010; 36: 33–50.

[20] Croley II TE. Laurentian Great Lakes double-Co2 climate change hydrological impacts. Climate Change. 1990; 17: 27–47.

[21] Croley II TE, Quinn FH: Great Lakes hydrology under transposed climates. Climate Change. 1998; 38: 405–433.

[22] Hayhoe K, VanDorn J, Croley II T, Schlegal N, Wuebbles D: Regional climate change projections for Chicago and the US Great Lakes. Journal of Great Lakes Research. 2010; 36: 7–21.

[23] Lee DH, Quinn FH, Sparks D, Rassam JC: Modification of great lakes regulation plans for simulation of maximum Lake Ontario outflows. Journal of Great Lakes Research. 1994; 20: 569–582.

[24] Lofgren BM, Quinn FH, Clites AH, Assel RA, Eberhardt AI, Luukkonen CL: Evaluation of potential impacts on Great Lakes water resources based on climate scenarios of two GCMs. Journal of Great Lakes Research. 2002; 28: 537–554.

[25] Mortsch LD, Quinn FH: Climate change scenarios for Great Lakes basin ecosystem studies. Limnology and Oceanography. 1996; 41: 903–911.

[26] Assel RA: The 1997 ENSO event and implication for North American Laurentian Great Lakes winter severity and ice cover. Geophysical Research Letters. 1998; 25: 1031–1033.

[27] Assel RA: Contemporary Lake Superior ice cover climatology. In: Munawar M, Munawar IF, editors. State of Lake Superior, Ecovision world Monograph series. Aquatic Ecosystem Health and Management Society, Canada, 2009: 66 p.

[28] Assel RA, Janowiak JE, Young S, Boyce D: Winter 1994 weather and ice conditions for the Laurentian Great Lakes. Bulletin of American Meteorology Society. 1996; 77: 71–88.

[29] Assel RA, Janowiak JE, Boyce D, O'Connors C, Quinn FH, Norton DC: Laurentian Great lakes ice and weather conditions for the 1998 El Nino winter. Bulletin of American Meteorology Society. 2000; 81: 703–717.

[30] Croley II TE, Lewis CFM: Warmer and drier climates that make terminal Great Lakes. Journal of Great Lakes Research. 2006; 32: 852–869.

[31] Angel JR: Cyclone climatology of the Great Lakes. IDEALS, Illinois State Water Survey, Publication MP-172. 1996; technical report.

[32] Lofgren BM: Simulated effects of idealized Laurentian Great Lakes on regional and large-scale climate. Journal of Climate. 1997; 10: 2847–2858.

[33] Logfren BM: A model for simulation of the climate and hydrology of the Great Lakes basin. Journal of Geophysical Research: Atmosphere. 2004; 109: D18108. doi: 10.1029/2004JD004602, 1–20.

[34] Biron S, Assani AA, Frenette J-J, Massicotte P. Comparison of Lake Ontario and St. Lawrence River hydrologic droughts and their relationship to climate indices. Water Resources Research. 2014; 50: 1396–1409, doi:10.1002/2012WR013441.

[35] Quinn FH: Lake Superior regulation effects. Water Resources Bulletin. 1978; 14: 1129–1142.

[36] Vanderpoorten A, Palm R: Canonical variables of aquatic bryophyte combinations for predicting water level trophic level. Hydrobiologia. 1998; 386: 85–93.

[37] Quessy J-F, Saïd M, Favre A-C: Multivariate Kendall's tau for change-point detection in copulas. Canadian Journal of Statistics. 2013; 41: 65–82.

[38] Lombard F: Rank tests for changepoint problems. Biometrika. 1987; 74: 615–624.

[39] Assani AA, Landais D, Mesfioui M, Matteau M: Relationship between the Atlantic Multidecadal oscillation index and variability of mean annual flows for catchments in the St. Lawrence watershed (Québec, Canada) during the past century. Hydrology Research. 2010; 41: 115–125.

[40] Burn DH: Climatic influence on streamflow timing in the headwaters of the Manckezie River basin. Journal of Hydrology. 2008; 352: 225–238.

[41] Enfield DB, Mestas-Nunez AM, TRimble PJ: The Atlantic multidecadal oscillation and its relation to rainfall and river flows in the continental US. Geophysical Research Letters. 2001; 28: 2077–2080.

[42] McCabe GJ, Palecki MA, Betancourt JL: Pacific and Atlantic Ocean influences on multidecadal drought frequency in the United States. Proceeding of the National Academy of Sciences of the USA. 2004; 101: 4136–4141.

[43] Sutton RT, Hodson DLR: Atlantic Ocean forcing of North American and European summer climate. Science. 2005; 309: 115–118.

[44] Tootle GA, Piechota TC, Singh A: Coupled oceanic-atmospheric variability and US streamflow. Water Resources Research. 2005; 41: W12408, 11 p.

[45] Bonsal BR, Shabbar A, Higuchi K: Impacts of low frequency variability modes on Canadian winter temperature. International Journal of Climatology. 2001; 21: 95–108.

[46] Bravets TP, Walvoord MA: Trends in streamflow in the Yukon River basin from 1944 to 2005 and the influence of the Pacific Decadal Oscillation. Journal of Hydrology. 2009; 371: 108–119.

[47] Gobena AK, Gan TY: Low-frequency variability in Southwestern Canadian streamflow: links with large-scale climate anomalies. International Journal of Climatology. 2006; 26: 1843–1869.

[48] Goodrish GB: Influence of the Pacific Decadal Oscillation on winter precipitation and drought during years of neutral ENSO in the western United States. Weather and Forecasting. 2007; 22: 116–124.

[49] Kim T-W, Valdés JB, Nijssen B, Roncayolo D: Quantification of linkages between large-scale climatic patterns and precipitation in the Colorado River basin. Journal of Hydrology. 2006; 321: 173–186.

[50] Mantua NJ, Hare SR: The Pacific decadal oscillation. Journal of Oceanography. 2002; 58: 35–44.

[51] McCabe GJ, Palecki MA: Multidecadal climate variability of global lands and oceans. International Journal of Climatology. 2006; 26: 849–865.

[52] Brown RD: Analysis of snow cover variability and change in Québec, 1948–2005. Hydrological Processes. 2010; 24: 1929–1954.

[53] Guerfi N, Assani AA, Mesfioui M, Kinnard C: Comparison of the temporal variability of winter daily extreme temperatures and precipitations in Southern Quebec (Canada) using the Lombard and copula methods. International Journal of Climatology. 2015; 35: 4237–4246.

[54] Lee DH, Quinn FH, Sparks D, Rassam JC: Modification of Great lakes regulation plans for simulation of maximum Lake Ontario outflows. Journal of Great Lakes Research. 1994; 20: 569–582.

[55] Quinn FH: Anthropogenic changes to Great Lakes water levels. Great Lakes Updates. 1999; 136: 1–4.

Chapter 3

Lake La Salada de Chiprana (NE Spain), an Example of an Athalassic Salt Lake in a Cultural Landscape

Rutger De Wit

Additional information is available at the end of the chapter

http://dx.doi.org/10.5772/6443

Abstract

On a global scale, athalassic inland salt lakes are abundant, albeit restricted to semiarid and arid climates. La Salada de Chiprana is unique in Western Europe, because it is a permanent and relatively deep (up to 5.6 m) hypersaline lake (40–90 g total dissolved salt L^{-1}) since 1700 AD. It forms part of a cultural landscape, which imposes a challenge for management. The aim of this paper is to describe the specific microbial biota and how they interacted with both animals and plant species during the last 25 years. The deeper parts regularly showed salinity stratification with an anoxic sulfide-rich hypolimnion and a bloom of green sulfur bacteria (*Prosthecochloris aestuarii* and *Chlorobium vibrioforme*) at the pycnocline. Despite highly eutrophic conditions, often the top water layer is transparent due to top-down control of phytoplankton populations by the brine shrimp, *Artemia parthenogenetica*. This allows for the development of submerged aquatic vegetation of the endemic foxtail stonewort *Lamprothamnium papulosum* var. *papulosum* f. *aragonense*, and microbial mat communities build by the cyanobacterium *Coleofasciculus (Microcoleus) chthonoplastes* coexisting with green filamentous nonsulfur bacteria (Chloroflexaceae). The microbial mats show photosynthetically induced precipitation of high-Mg calcite, which by incorporating viruses represents a mechanism for their fossilization.

Keywords: microbial mat, multiple stable states, regime shift, charophyta, cyanobacteria, *Coleofasciculus (Microcoleus) chthonoplastes*, Chloroflexaceae

1. Introduction

Saline aquatic environments on the Earth include the ocean and the transitional environments (lagoons, estuaries, coastal wetlands, river mouths, etc.) on the coast where the salinity is derived from the sea, and epicontinental streams and lakes without a recent connection to the sea. The former are called thalassic (i.e., derived from the sea) and the latter athalassic. On a global scale, athalassic inland salt lakes are widely distributed (see **Figure 1**). These areas are characterized by their arid or semiarid climate, with annual rainfall values of 25–200 and 200–500 mm, respectively, which are largely insufficient to compensate for evaporation. In a landscape setting, such a structural water deficit does not allow the formation of permanent rivers that slow down the formation of river valleys. In many cases, the water simply flows to depressions in the landscape from where it escapes to the atmosphere by evaporation. Such inland areas that drain to a depression in the landscape with no outlet to rivers flowing to the sea are called endorheic basins. Permanent and temporal salt lakes are typically found in these endorheic basins where they often occupy the lower parts. In addition, athalassic salt lakes occur in Antarctica, where high salinities in the lake are a consequence of outfreezing. During freezing of water, ice crystals of almost pure water are formed while the dissolved salts are expelled from the freezing volume to the remaining water that becomes progressively more concentrated in total dissolved salts. Lake Bonney (77°43′S 162°22′E) is a permanent ice-covered athalassic saline lake on the Antarctic continent in the McMurdo Dry Valleys. On a global scale, inland saline lakes represent a volume of 85 × 103 km^3, which is not much lower than the global volume of freshwater lakes of 105 × 103 km^3 (Shiklomanov 1990 cited by Williams [1]). However, the Caspian Sea represents almost 75% of the total volume of inland salt lakes. Other well-known salt lakes include the Dead Sea (Middle East), the Aral Sea (Central Asia), Mono Lake in California (USA), Mar Chiquita (Argentina), Sambhar Salt Lake (India), and Lake Eyre (Australia). In Western Europe, Spain is the only country where inland athalassic saline lakes occur in three different regions. The majority of these saline lakes are ephemeral showing alternations between shallow water bodies during wet periods and dry salt flats during dry periods. Lake La Salada de Chiprana is a unique salt lake in Western Europe because it is permanent and deep (up to 5.5 m depth).

Figure 1. Global distribution of athalassic salt lakes (shaded areas indicate the regions where such lakes occur naturally) from Williams [1]. With kind permission of Cambridge University Press, Environmental Conservation.

Limnology is the scientific study of the physical, chemical, geological, and ecological dynamics of inland lakes and other inland water bodies. The term was introduced by François-Alphonse Forel (1841–1912), who studied Lake Geneva. Most limnologists neglected the study of inland salt lakes for a long time, despite their global importance. Even today, the salt lakes do not receive the interest they merit as it is reflected by the fact that limnology is sometimes incorrectly described as the discipline studying freshwater systems. The original focus on freshwater lakes in limnology can be explained by historic reasons as this discipline really started to develop in Western Europe around the turn of the twentieth century, where inland salt lakes are scarce as mentioned above. Even around the 1920s, limnologists were mainly active in Western and Central Europe as well as in the eastern part of the USA where freshwater lakes predominate. By the end of the 1940s, Ramón Margalef (1919–2004) initiated the study of the inland Spanish salt lakes and American and Australian researchers became interested in the salt lakes in the semiarid and arid parts of their large countries. William David Williams (1936–2002) has been particularly active in promoting the study of salt lakes at a global level and highlighted the threats to inland salt lakes and advocated the need for conservation measures [2].

The salinity is highly variable among athalassic saline lakes both in terms of concentration and composition of dissolved salts. The salt often derives from the soils in the endorheic basin, which are lixiviated by rainwater that carries the dissolved salts to the terminal lake. Hence, the salt composition of the inflowing water is strongly determined by the parent material and the sedimentary deposits that form the soils in the endorheic basin. The concentration of salts in salt lakes is generally expressed in units of total dissolved salts per liter (TDS L^{-1}). Inland lakes are considered as saline lakes when the concentration of total dissolved salts exceeds 3 g TDS L^{-1}. However, the lakes that have a salinity exceeding 10 g TDS L^{-1} are home to biotas that are clearly different from that of freshwater lakes. Inland lakes with salinities between 3 and 10 g L^{-1} are often referred to as subsaline. On the other side, lakes with salinities exceeding 40 g TDS L^{-1} are considered as hypersaline. Dissolved salts in inland lakes mainly include the anions carbonates (CO_3^{2-} and HCO_3^-), chloride (Cl^-), sulfate (SO_4^{2-}), and silicate ($SiO(OH)_3^-$ and other forms), and the cations sodium (Na^+), magnesium (Mg^{2+}), calcium (Ca^{2+}), and potassium (K^+). The final composition in the lake water depends on both, the composition of the inflowing water and the mineral precipitation processes. Hence, high Ca^{2+} and low Mg^{2+} give rise to lakes dominated by halite (NaCl), while high Ca^{2+} combined with high Mg^{2+} gives rise to Na^+, Mg^{2+}, Cl^- and often SO_4^{2-} dominated lakes ranging in pH from 6 to 8.5. Lake Chiprana is an example of the latter and is characterized by a high pH for this group, i.e., a pH around 8.5. Interestingly, low Ca^{2+} and low Mg^{2+} give rise to so-called soda lakes dominated by Na^+ and CO_3^{2+} and a high pH of 10–11 [3]. A precise determination of TDS requires a chemical analysis of the major ions. Complete evaporation of saline water and gravimetric determination of precipitated salt are inadequate for TDS determinations, because water molecules are incorporated in different crystals during the precipitation. Moreover, the use of a measure of conductivity may be problematic. At first, the relation between TDS and conductivity depends on the composition of dissolved salts and the relationship is, therefore, different among lakes. Second, the relation between TDS and conductivity breaks down at higher salinities, particularly when bivalent ions are abundant in solution [4].

Inland saline lakes in a pristine landscape are ideal sentinels for variations and changes in the climate. Hence, in permanent salt lakes, draught periods are reflected by the lowering of the water table, decreasing water volume, and increasing salinity as well as a decreasing surface area. The latter is due to exposure of the shallowest parts. From the littoral zone into the lake, the shallow areas are designated as playas. These playas typically alternate between emersion after dry periods and submersion following wet periods. The exposed bottom of playas is often covered by a salt crust. However, prolonged exposure for many years may result in colonization by halophyte vegetation. During wet periods, the reverse occurs. These variations can potentially be monitored by remote sensing. For temporal salt lakes, increasing draught results in shorter submersion and longer exposure where the lake bottom appears as a salt pan. For example, in the temporal lake La Dehesilla (Spain), during the water years 1989 and 1990, the lake was filled with water during 5 months and completely dry during 7 months. The salinity of the water varied from 20 g TDS L^{-1} to 90 g TDS L^{-1} during the wet periods [5]. Potentially, the duration of the submerged and exposed periods can be monitored by remote sensing.

Nevertheless, humans nowadays heavily impact the inland salt lakes and jeopardize the conservation of the unique values of these ecosystems [1]. The human threats include inflow diversions, irrigation, mining, pollution, and biological disturbances, e.g., the introduction of exotic species. In many salt lakes, the variations of water volume and salinity do no longer faithfully reflect the natural climatic variation alone, but are rather modified by many of these human impacts. There is an urgent need for conservation measures that are particularly adapted to inland salt lakes. A good example for conservation measure is Mono Lake in California [1]. In this chapter, I describe some of the unique values of the athalassic salt lake La Salada de Chiprana, with a particular focus on the microbial communities. In addition, I discuss the management actions put in place for the conservation of this lake. However, Chiprana Lake is not a pristine saline lake as humans have heavily impacted the catchment since 1700. For centuries, this allowed the existence of this permanent lake in a cultural setting. Major changes in agriculture since the 1950s have represented a threat to the unique values of this lake. This poses the question on how to develop conservation measures for this interesting salt lake in a changing cultural context.

2. Lake La Salada de Chiprana as a specific case of an athalassic salt lake in a cultural landscape

2.1. Localization and history of lake La Salada de Chiprana

"La Salada de Chiprana" (41°14′30″N, 0°10′50″W, NE Spain) is the only permanent athalassic shallow salt lake in Western Europe. This salt lake (**Figure 2** and **3**) is located in an endorheic basin in the Ebro Valley 6 km SW of the village of Chiprana, and occupied a total surface of 31.5 ha in 1991 and varied in surface between 36 ha in 1993 and 22 ha in 2008. Today, the lake La Salada de Chiprana is a protected area which benefits from the following qualifications: (1) Ramsar wetland site since 1994, (2) site of community interest (SCI) according to the Habitat

Directive of the European Union since 1997, and (3) natural reserve of the Gobierno de Aragón since 2006.

Figure 2. Lake La Salada de Chiprana (Aragón, Spain) located in NE Spain in the valley of the Ebro River close to the Mequinenza Reservoir (41°14′30″N, 0°10′50″W; see maps). Panoramic view of lake La Salada de Chiprana from the southwestern shore of the lake showing palaeocanals and playas. Close-up of paleocanal with playa in the back.

This salt lake lies upon the upper Oligocene-Miocene Caspe formation that mainly comprises sand and siltstones deposited by a river. Hence, the area where the salt lake occurs is located in a former river valley, where the river meandered. Gravel and sand were being deposited in the riverbed, while silt was deposited in the floodplain during flooding of the river. The differences in compaction and erodability of the sand and siltstones have resulted in an inversion of topography in the landscape, with the meandering sandstone formations emerging as ridges in the landscape. These sandstone formations are, therefore, called paleocanals (see **Figure 2**).

A paleolimnological study [6] has shown that during most of the Mid and Late Holocene, lake La Salada de Chiprana was a temporal lake. Human activities in the surroundings included deforestation, irrigation, and olive tree planting. In the thirteenth century, a small irrigation canal, the Civán canal, was created to carry water from the *Guadalupe* River into the surroundings of the lake. All these changes resulted in increased flow of water into the lake and it thus became a permanent hypersaline lake since 1700 AD. Irrigation practice was formally organized since 1838, when a committee of 40 landowners was formed. The irrigated surface reached 3912 ha during the 1950s and was reduced by 400 ha because of the construction of a dam and the large Reservoir of Mequinenza in the Ebro River in 1967. In 1988, a smaller reservoir was built in the *Guadalupe* River to allow for an increase of irrigation in the region. A new Civán canal was built, running parallel to the old canal, which obtains its water directly from this reservoir. Since then, an additional 200 ha has been irrigated. Olive trees were the main crops until 1970. Nowadays, most of the crops include cereals and alfalfa [6]. Hence, for the last 300 years, lake La Salada de Chiprana has been subjected to both climatic and anthropogenic influences. The watershed *sensu stricto* of the lake is of 768 ha of which 515.7 ha benefit from protection measures, i.e., 154.8 ha as a natural reserve and 360.9 ha as a peripheral buffer zone. However, due to the irrigation canals, the watershed receives water inputs from the outside of this watershed. In the early 1990s, scientists expressed their concern about the conservation of the unique properties of the lake and called for regulation of the water flows in the area.

2.2. Regulating the water flows and management of the Salada de Chiprana Lake

Irrigation water for agriculture in the watershed of the Salada de Chiprana is drawn from the reservoir in the *Guadalupe* River and brought to the watershed of the lake via open irrigation canals. These comprise the Civán canal that connects to the smaller Acequia de Farol and the Acequia de la Casilla canal in the surroundings of the Salada de Chiprana. Moreover, overflow of an upstream irrigation zone also feeds into the Acequia de Farol and the Acequia de la Casilla, thus, representing an additional input. Often, more water is channeled to the watershed of the lake than requested for irrigation. Before 1994, this excess irrigation water was directly channeled into the Salada de Chiprana Lake. This practice created major environmental problems (see below) and, therefore, it was decided to build a dam, which allowed diverting this surplus flow via the Arroyo del Regallo River into the Reservoir of Mequinenza [6, 7]. In 1993, more than 100,000 m^3 freshwater from Civán canal entered the lake in 1993 [6]. The hydraulic works have been used since 1994. In addition, measures were taken to prevent the water from a small irrigation canal, i.e., the small Acequia de la Casilla, entered the lake directly [7].

Between 1989 and 1993, the lake level rose due to dumping of surplus irrigation water and irrigation returns in the watershed. As a result, by 1993, the lake salinity had been reduced to 30–40 g TDS L^{-1} and the surface area of the lake achieved a maximum of 36 ha. Between January 1994 and September 1995, the lake level decreased because the diversion and salinities increased again to 80 g TDS L^{-1}. Between 1995 and 1999, the lake level rose again, while after 2000, there was a persistent trend of decreasing lake levels. In the early 1990s, the maximum depth reported for the lake was 5.6 m [8]. Between the early 1990s and early 2009, the lake level decreased overall by 1.75 m [9]. Hence, on 4 August 2007, the Confederación Hidrográfica del Ebro recorded a maximum depth of 3.7 m. The surface areas and volumes of the salt lake have been calculated from the topography of the landscape and the level and bathymetry of the lake [9, 10]. Superimposed on these trends, since 1998, the lake also showed repeatable intra-annual fluctuations with amplitude of 30–50 cm [9]. Such behavior is natural for these salt lakes as water levels rise during winter and spring and decrease during summer, thus achieving their annual minimum levels in late summer or autumn. The decreasing lake level between 1991 and 2008, corresponded to a 42% reduction in surface (from 36 to 21 ha) a 55 % reduction in volume (from 893,000 to 405,000 m^3) [9]. The decrease in the surface is illustrated in **Figure 3**. In accordance with the lake level, the salinity increased from 40 to 90 g TDS L^{-1} (**Figure 3**). Nevertheless, the ionic composition remained remarkably constant during the entire 17 years and was dominated by Na^+, Mg^{2+}, SO_4^{2-}, and Cl^-. The molar ratios of dissolved ions (n = 53) were 1.13 ± 0.24 SO_4^{2-}/Cl^-, 1.38 ± 0.34 Na^+/Cl^-, and 0.97 ± 0.19 Mg^{2+}/Cl^-. Magnesium exceeded calcium showing more variable molar ratios for Mg^{2+}/Ca^{2+}, i.e., 21.4 ± 10.5 (n = 53). The total amount of salts in the lake was estimated to be equal to 44,750 tons. After 1995, when the water level dropped below 137.6 m (European Datum georeference system for W Europe, ED50) above mean sea level (Alicante), a small basin in the SE corner became separated from the main lake (**Figure 3**). This separated 822 tons of dissolved salt representing a loss of 2% of salt for the main lake. Nevertheless, the salinities in the lake showed remarkable conservative behavior, i.e., the mass of total dissolved salt in the lake was rather constant and variations of its salinities were directly related to the volume of the lake [9].

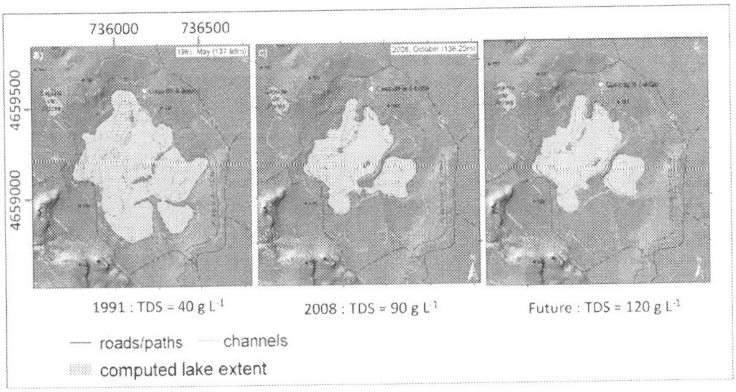

Figure 3. Lake extent of the lake La Salada de Chiprana and updated bathymetry calculated from the bathymetric map of Guerrero et al. [10] and based on monitoring of water height at a gauge and cartography (IGN Spain), adapted from De Wit et al. [9], with kind permission of John Wiley. TDS = concentration of total dissolved salts. The right panel contains a prediction for a future extent when salinity would have increased to 120 g TDS L^{-1} and that can be expected to occur before 2020, if the trend initiated since 2000 persists.

In conclusion, after a period of excessive water discharges in the lake in the early 1990s, since the introduction of the measures in 1993, the lake hydrology has been governed by high evapotranspiration (1000–1500 mm $year^{-1}$), low rainfall (200–400 mm $year^{-1}$), water runoff in its endorheic watershed, irrigation returns and groundwater flows. This resulted, however, in persistently decreasing lake levels and increasing salinities since 2000. To counteract this trend, the managers have decided in 2013 to supplement clean freshwater into the lake from the Civán River.

2.3. Variations in microbiological communities during the last 25 years in La Salada de Chiprana

2.3.1. Planktonic communities and stratification of the water column

During the early 1990s, when the lake contained its highest amount of water and showed some parts where depths reached more than 5 m, stratification of the water column was often observed [8]. **Figure 4** describes the stratification that I observed on 24 January 1990. The stratification was based on a salt gradient. The upper water layer had been diluted by the freshwater inflow and showed lower salinities than the bottom layer. The top layer or epilimnion was separated from the bottom layer (hypolimnion) by a density gradient or pycnocline. Temperature also showed a gradient from 6°C in the epilimnion to 17°C at 2 m depth. This is surprising as distilled water has a maximum density at 4°C and above 4°C its density decreases with temperature. One would, thus, expect that the water of 6°C sinks to the bottom in a lake where temperature achieves 17°C. Nevertheless, this was not the case as the cold epilimnion water had a higher density than the warmer hypolimnion, due to the very strong differences in salt concentrations. Temperature effects counterbalanced the salt gradient but were clearly not sufficient to destroy the stratification. This phenomenon is known as an inverse temperature gradient and can, sometimes, be observed in hypersaline systems. The highly saline water has a strong tendency to absorb the energy from solar radiation. In addition, the heat capacity of water decreases with increasing salinity and, thus, for the same amount of energy absorbed, saline water increases more strongly in temperature than freshwater.

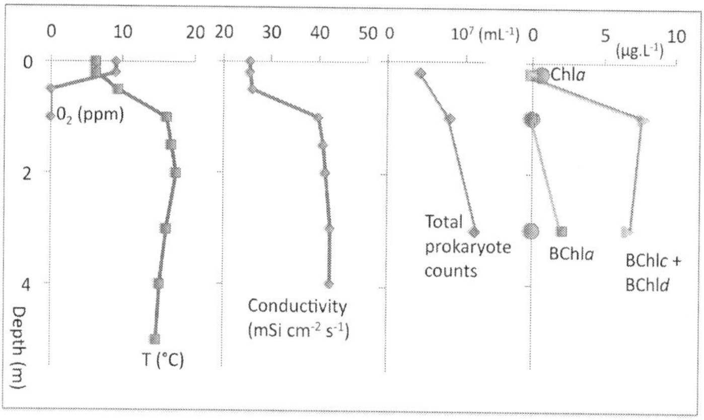

Figure 4. Stratification of physicochemical and microbiological variables in the deepest part (5.5 m) of the lake observed on 24 January 1990. The stratification is based on a gradient of salt concentrations (pycnocline), which is reflected by conductivity. The upper part of 50 cm thickness (epilimnion) is oxic and low concentrations of Chlorophyll *a* (Chl*a*) reflect the presence of microalgae. Green sulfur bacteria, which are strict anaerobes, and phototrophic bacteria develop in the anoxic zone at the pycnocline. These organisms contain bacteriochlorophylls *c* and *d* (BChl*c* and BCl*d*) as light-harvesting pigments, together with minor proportions of bacteriochlorophyll *a*.

The bottom layer and the pycnocline were completely anoxic, while only the epilimnion was oxic. A low density of phytoplankton was reflected by a low chlorophyll *a* content of 1 µg L^{-1}. At the pycnocline, still sufficient light was present to favor the blooming of phototrophic bacteria, i.e., green sulfur bacteria. These bacteria use H_2S as the electron donor in anoxygenic photosynthesis. Morphological observations with light microscopy indicated that cells similar to *Chlorobium vibrioforme* were dominant, although they often coexisted with minor populations of cells resembling *Prosthecochloris aestuarii*. The latter was often located deeper in the lake. Xavier Vila and his colleagues at the University of Girona (NE Spain) succeeded in isolating these green sulfur bacteria and obtained strains in pure culture, which were related to *Prosthecochloris aestuarii* SK413 and *Chlorobium vibrioforme* DSM260 [11]. Green sulfur bacteria contain chlorosomes with bacteriochlorophylls capable to efficiently harvest light at low intensities. Both isolated strains contained different allomers of bacteriochlorophylls *c* and *d* (BChl*c* and BChl*d*) and the carotenoid chlorobactene [11]. In addition, the green sulfur bacteria also contain minor amounts of bacteriochlorophyll *a* (BChl*a*) that occurs (i) in the photosynthetic reaction centers, (ii) in the baseplate of the chlorosomes, and (iii) as associated with the Fenna-Matthews-Olsen protein, all involved in the photosynthetic light processing. Hence, the BChl*a* measured has probable originated from these green sulfur bacteria in the lake.

Interestingly, the sequencing of the 16S rRNA genes and the fmoA gene, the gene coding for the Fenna-Matthews-Olsen protein, indicated that specific lineages of green sulfur bacteria exist, which are adapted to marine and saline environments. Nonetheless, within such a lineage, the *Prosthecochloris* strain isolated from Chiprana was actually quite different from the type strain of *Prosthecochloris aestuarii*. The *Prosthecochloris* strain from Chiprana clustered with a strain isolated from another inland salt lake in the Death Valley indicating that these inland saltwater lake bacteria were separated from their marine relatives early during evolution [12].

The strong variations in salinities observed in the lake since the 1990s, triggered important changes in the water column communities. In 1990, the Chl*a* concentrations in the top layer were low due to a strong top-down effect of the water column grazer, the brine shrimp *Artemia parthenogenetica* (Bowen & Sterling, 1978). During 1992 and 1993, the strong input of freshwater resulted in decreasing salinities with a wane of the brine shrimp populations as several species in this genus do not occur below 35 g TDS L^{-1} [13]. Hence, in the absence of the top-down control by brine shrimps, phytoplankton was able to develop high densities (with Chl*a* values up to 100 $\mu g\ L^{-1}$), which was probably further enhanced by the high nutrient input into the lake from the surplus irrigation water [14]. Fortunately, after 1994, with increasing salinities the brine shrimps returned to the lake and the top-down control was reestablished. The decreasing lake level and depth also resulted in less frequent occurrence of stratified periods. This area is subjected to strong winds that act on the lake surface to induce strong mixing in the lake and destruction of the pycnocline. In addition, the reduced freshwater input from the watershed also implies that the top layer of the lake is exposed to a lesser degree of dilution than before, which slows down the buildup of a pycnocline.

2.3.2. Benthic communities and their variations

Between 1989 and 1991, a large part the submerged playas were covered by benthic microbial mats [8] build by the cosmopolitan cyanobacterium *Coleofasciculus chthonoplastes* (Thuret ex Gomont) Siegesmund, Johansen et Friedl 2008. This is a filamentous cyanobacterium that forms bundles in a common sheath (see **Figure 5A**). The species was previously known as *Microcoleus chthonoplastes*, but a detailed phylogenetic study showed that among the filamentous *Microcoleus* species, the marine and halotolerant species including *M. chthonoplastes* formed a lineage that was clearly different from the freshwater *Microcoleus* branch; therefore, the species has been renamed by Siegesmund et al. [15]. Flowering plants (Magnoliophyta) were also observed as *Ruppia* sp. that formed clear meadows. The development of *Ruppia* meadows is clearly antagonistic to the microbial mats as the plants occupy space, modify the sediment surface and reduce light intensities at the sediment surface by shading of the canopy. During this period, the extension of the *Ruppia* meadows, a phenomenon that was attributed to decreasing salinities [8], resulted locally in replacement of the *C. chthonoplastes* mats.

Diaz et al. [14] observed that the continuously decreasing salinities between 1990 and 1993 in combination with increased eutrophication resulted in heavy epiphytic growth on *Ruppia*, often inducing their decline. Simultaneously, the microbial mats continued their demise, probably because of reducing light at the sediment surface due to increasing shading by phytoplankton (see Section 4.1). The excessive low salinities in the lake, together with the progressive disappearance of the microbial mats and the poor quality of the *Ruppia* meadows, motivated Spanish researcher to ring the bell and call for targeted management measures that resulted in the hydraulic works achieved in May 1993 [7] (see Section 3).

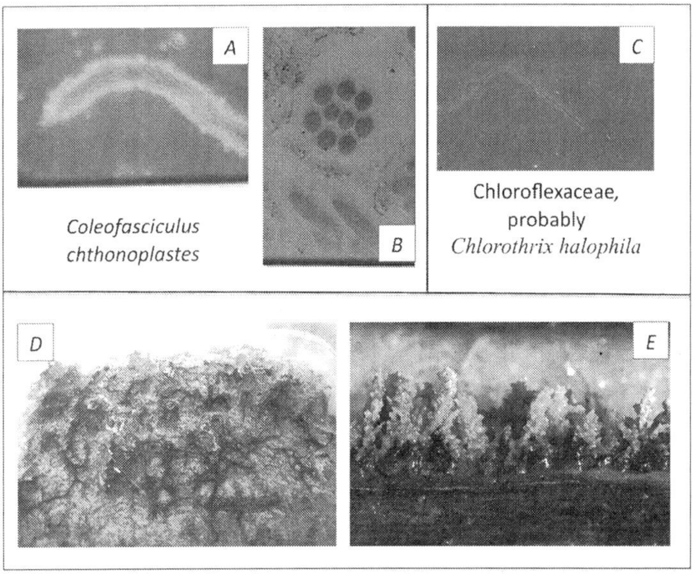

Figure 5. (A) Phase contrast light microscopic image of *Coleofasciculus chthonoplastes* (Thuret ex Gomont) Siegesmund, Johansen et Friedl 2008 (formerly *Microcoleus chthonoplastes*) showing multiple filaments in a common bundle. (B) Transmission electron microscopic image of a cross section of a bundle of *Coleofasciculus chthonoplastes.* (C) Phase contrast light microscopic image of a filamentous green nonsulfur bacterium from the Chloroflexaceae family, probably *Chlorothrix halophila.* (D) Macroscopic image of the surface of a biofilm growing on the top of the debris of the foxtail stonewort (*Lamprothamnium papulosum* var. *papulosum* f. *aragonense* (Prósper) Wood). The biofilm was mainly composed of green nonsulfur bacteria from the Chloroflexaceae family. (E) Macroscopic image (macrophotography) of a cross section of a mat build by *Coleofasciculus chthonoplastes* in 2010 under very high salinities (>90 g TDS L^{-1}). The purple layer indicates the development of purple sulfur bacteria in these mats. Note a fine layer of fine-grained high-Mg calcite crystals at different depths in the mat. Due to high salinities, gypsum started to precipitate on the surface of the mats where it formed beautiful crystals.

The situation was clearly reversed after the inauguration of these hydraulic works. Valero-Garcés et al. [6] reported that since 1995, a particularly dry year in this region (200 mm rainfall), water column transparencies had recovered in the lake. During a visit in September 1996, I observed reestablishment of *C. chthonoplastes* microbial mats. Recolonization by microbial mats continued and by 1999, the extent of microbial mats was similar to that described for the lake in the late 1980s and early 1990s [8]. However, the *Ruppia* meadows had completely disappeared by 1996 as a result of the increased salinities.

More detailed studies by Jonkers et al. [16] described the laminated structure of these mats as multilayered microbial mats build by *C. chthonoplastes* living in association with filamentous green nonsulfur bacteria from the Chloroflexaceae family. Chloroflexaceae show thin filament morphologies of light green color, which do not fluoresce red light upon blue excitation and can, thus, be distinguished under the microscope from the thin cyanobacteria filaments, e.g., *Leptolyngbya*. As green sulfur bacteria (see **Figure 4**), the filamentous green nonsulfur bacteria also contain BChl*c* and BChl*d*. These bacteriochlorophylls are involved in light harvesting and are special, because rather than representing a single macromolecule, they comprise mixtures of molecules known as allomers that may differ by the alcohol esterified to the macrocycle and substituents on the macrocycle. The different allomers can be separated by HPLC, using a reverse-phase protocol as that described by Buffan-Dubau et al. [17] (see **Figure 6**). The phylogenetic study by Bachar et al. [18] has confirmed the presence of 16S rRNA gene sequences affiliated to the Chloroflexaceae family some of them very closely related to the Candidatus *Chlorothrix halophila* (see **Figure 5C**). Multiple white layers of calcium carbonate were observed in the mat (see **Figure 5E**), which have been identified as high-Mg calcite [19]. In addition, diatoms sometimes occurred in the mat, particularly on the top surface layer, particularly during spring. These diatoms are favored by high inputs of inorganic nitrogen, i.e., NO_3^- and NH_{4+} [20]. In contrast, the cyanobacteria were stimulated by the combined addition of inorganic nitrogen end phosphate [20].

The microbial mats fixed N_2 [21]. In general, *C. chthonoplastes* does not possess the genes for N_2 fixation, although one strain of this species was shown to contain these genes, which it probably obtained through lateral gene transfer [22]. Hence, it appears most likely that the N_2 fixation is attributed to the heterotrophic bacteria in the mat [21], although N_2 fixation by cyanobacteria cannot be excluded [22].

Between 2000 and 2006, there was a spectacular development of charophytes, i.e., a specific form of the foxtail stonewort *Lamprothamnium papulosum* var. *papulosum* f. *aragonense* (Prósper) Wood. This charophyte has been described originally as a distinct species by Eduardo Reyes Prósper in 1910 as a typical plant for the semiarid steppe lakes in Aragon [23, 24]. This form of *Lamprothamnium* is able to thrive at extremely high salinities (70–90 g TDS L^{-1}). In 2000–2001, sparse populations coexisted with the *C. chthonoplastes* microbial mats. Some of the larger individuals were overgrown by slimy biofilms comprising diatoms and Chloroflexaceae, which at the sediment interface coalesced with the microbial mats. However, from 2002, this charophyte dramatically increased its surface and at many places supplanted the existing *C. chthonoplastes* microbial mats. This colonization culminated during 2005 in extensive prairies covering virtually all sediments between 0.2 and 1.8 m water depths.

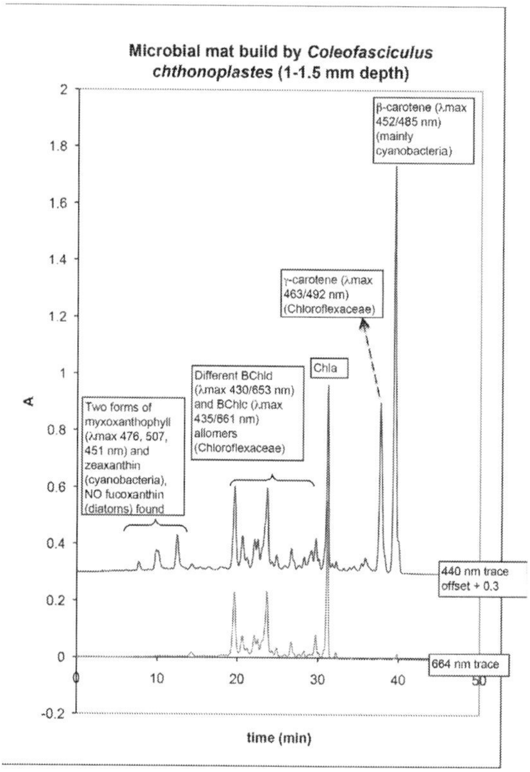

Figure 6. HPLC traces of lipophilic pigments extracted from the 1–1.5 mm depth layer of a microbial mat build by the cyanobacterium *Coleofasciculus chthonoplastes* associated with Chloroflexaceae. The pigment analysis was performed after the protocol described by Buffan-Dubau et al. [17] using a diode array detector (350–800 nm). By comparing the chromatograms selected at 440 and 664 nm, we can directly see whether the peaks correspond to a carotenoid (no absorption at 664 nm) or to a green chlorophyll (Chl*a*) and bacteriochlorophylls (BChl*c* and BChl*d*). Note the different peaks of BChl*c* and BChl*d*, which correspond to different allomers that may differ by the alcohol esterified to the macrocycle and substituents on the macrocycle. λ_{max} gives the absorption maxima of the separated pigments measured in-line with the diode array detector.

In September 2006, it was observed that the shallower *L. papulosum* populations had become exposed to air [9] as the water level had further decreased by 30 cm since September 2005. Full sunlight and drying out of the top part triggered massive decay of *Lamprothamnium* and detritus of this charophyte accumulated in the lake. In the shallower parts of the lake, photosynthetic biofilms developed on top of this *Lamprothamnium* detritus (see **Figure 5D**). These biofilms were dominated by Chloroflexaceae together with different species of filamentous cyanobacteria. The remaining multilayered microbial mats build by *C. chthonoplastes* and Chloroflexaceae that had been restricted in area in 2005 to the littoral zone up to a depth of 20 cm, were now completely emerged and dried out. Hence, healthy *C. chthonoplastes* mats had almost completely disappeared from the lake. The biofilms, mainly comprising Chloroflexaceae, persisted until the spring of 2007. After an almost complete wane of the *C. chthonoplastes* builds microbial mats in 2006, there was again a new colonization by young mats of the shallower parts of the lake since the summer of 2007. However, during this period, salinities continued to increase. This resulted, in 2010, in the precipitation of gypsum ($CaSO_4$) at the surface of the newly established multilayered microbial mats build by *C. chthonoplastes* (see **Figure 5E**). During the 25-year-period, the microbial mats build by *C. chthonoplastes* waxed and waned, and this chronology has been described in more detail by De Wit et al. [9].

2.3.3. Interactions between the pelagos and the benthos

When combining the observations about the pelagos and the benthos, it appears that regime shifts are a characteristic feature in La Salada de Chiprana and may, thus, occur commonly in hypersaline lakes. Hence, we can clearly discern different forms of an ecosystem state characterized by low water turbidity, low phytoplankton and strong development of benthic communities. This low turbidity and low phytoplankton state seems to be obtained via top-down control of phytoplankton by the brine shrimp *Artemia parthenogenetica*. Within these benthic communities, the microbial mats build by *C. chthonoplastes* associated with Chloroflexaceae occupy a conspicuous role. In some cases, these mats completely dominate the benthic community in the lake, particularly at the shallower areas. Nonetheless, macrophytes may enter in competition with these microbial mats for space at the lake bottom. At lower salinities, probably up to 45 g TDS L^{-1}, the microbial mats may suffer from competition with *Ruppia* species. At higher salinities, the microbial mats may enter in competition with the extremely halotolerant foxtail stonewort *Lamprothamnium papulosum* var. *papulosum* f. *aragonense* (Prósper) Wood. Nevertheless, the observations after 2007 indicate that this charophyte cannot colonize the lake when salinities are above 90 g TDS L^{-1}.

Hypersaline systems are often, by nature, eutrophic. The athalassic lakes represent the terminal for runoff and groundwater flows that carry both dissolved salt and nutrients. The water budget is equilibrated by high evaporation from the lake. As a result, salt lakes accumulate dissolved salts and nutrients. Therefore, in the long-term, they tend to become eutrophic and more saline. During the low turbidity and low phytoplankton state, most of the nutrients will be captured by the benthic communities, i.e., the microbial mats or the macrophytes, both Magnoliophyta as *Ruppia* sp. and charophytes as *Lamprothamnium papulosum*. However, when the top-down control is being relieved due to disappearance of brine shrimps from the lake, the fast growing phytoplankton may quickly build up high densities and the water becomes turbid. The increasing turbidity will have a negative impact on the performance of the benthic mats and macrophytes. As a result, more and more nutrients will become available for the phytoplankton and the system may flip over into a turbid state as it had been observed in 1993 and 1994 by Diaz et al. [14]. Such an interaction scheme is characterized by a positive feedback loop for phytoplankton, i.e., the development of phytoplankton decreases the competitive abilities of the benthic communities and, thus, favors its own growth. This poses the question whether multiple stable states do exist for hypersaline lakes as these have been described for freshwater lakes by Scheffer et al. [25], for which the alternate states are characterized by (1) phytoplankton-rich high turbidity without submerged benthic vegetation and (2) highly transparent (clear) water, low phytoplankton with extensive meadows of submerged aquatic vegetation, respectively. Nevertheless, to prove the presence of multiple stable states, it is not sufficient to document regime shifts, but rather it needs to be shown that for a given window of environmental conditions, two stable states do potentially exist, each of them stabilized by positive feedback loops.

To define the windows for the different ecosystem states in this hypersaline lake, it appears most important to consider the ecology of the brine shrimp, because of its role in top-down control. Brine shrimps from different geographic origin have been reported to survive at salinities ranging from 35 to 110 g TDS L^{-1}. Assuming that this salinity range also applies for the growth and survival of the species in the lake *Artemia parthenogenetica*, we can expect that in this salinity window, a highly transparent water layer will favor benthic communities, i.e., most often microbial mats build by *C. chthonoplastes* co-occurring with *Ruppia* and/or *Lamprothamnion*. When moving outside this window of salinities, we expect regime shifts. This has, indeed, been observed during decreasing salinities in the early 1990s and I expect it to happen in the future, if the water level continues to drop down. However, both shifts do not proof the existence of multiple stable states. Multiple stable states could exist in the salinity window ranging from 35 to 110 g TDS L^{-1}, if certain conditions exist that keep the densities of the brine shrimp very low. However, I am not aware of an efficient top-down control mechanism that could keep the brine shrimps in check as the high salinities prevent the occurrence of fish or invertebrate predators.

2.4. *Coleofasciculus chthonoplastes* microbial mats as stromatolites

The laminated structure of the *C. chthonoplastes* (see **Figure 5E**) shows that this benthic microbial mat system can be considered a living stromatolite, which is a layered benthic microbial community that accretes due to the trapping, binding and cementation of sedimentary grains. These mineral grains incorporated in the organic matrix can be derived from the substratum, from external inputs or through the precipitation of mineral particles. The biofilms show the clear presence of fine layers of carbonate grains (**Figure 5E**), which turned out to be high-Mg calcite [19]. Ludwig et al. [26] reported that the precipitation of these grains was driven by photosynthesis that results in higher pH values, locally, in the mat. Hence, the precipitation of high-Mg calcite can be described by the following reaction equation: $Mg^{2+} + 4 Ca^{2+} + 5HCO_3^- + 5 OH^- \rightarrow MgCa_4(CO_3)_5 + 5H_2O$. This represents the stoichiometry for a high-Mg calcite crystal with 20% (molar) proportion of Mg. In practice, the values of Mg content in the high-Mg calcite are variable [19].

Viruses are engulfed into these high-Mg calcite crystals, which represents a mechanism for the fossilization of viruses and their long-term conservation in the geological record. This was shown recently by using three different microscopy techniques (epifluorescence, electronic and atomic force microscopes). These viruses were 50–80 nm in diameter, while energy-dispersive X-ray spectrometer analysis showed that they contain nitrogen and phosphorus in a molar ratio of approximately 9, which is typical for viruses of that size. Nucleic acid staining revealed that they contain DNA or RNA. As characteristic for hypersaline environments, the concentrations of free and attached viruses were very high (more than 10^{10} viruses per g of mat). Acid treatment that dissolved the high-Mg calcite showed that there were approximately $15 \rightarrow 10^{9}$ viruses per g of high-Mg calcite [19].

The observation in lake La Salada de Chiprana are also interesting in the way that it gives additional ideas why the extension of stromatolites has decreased dramatically during geological times. The current paradigm is that the evolution of metazoans has resulted in increased grazing on the microbial communities of the stromatolites and that, therefore, the present stromatolites are restricted to extreme environments where grazing pressure is low. Such extreme environments include the hypersaline environments. The observations show that the *C. chthonoplastes* build microbial mats can also be exposed to competition for space with macrophytes, which includes both the Magnoliophyta *Ruppia* sp. as well as the charophyte *Lamprothamnium papulosum* var. *papulosum*.

3. Conclusions and perspectives

Athalassic saline lakes are home to interesting biota, and their conservation is most important for scientific research, potential economic exploitation and aesthetics. When located in a pristine landscape, these lakes are superb sentinels for detecting climatic variations through monitoring of lake level, extent, salinities and biota. However, on a worldwide scale, many athalassic salt lakes have suffered severe degradation and it has been predicted that during this decade, the surface and volume of these lakes will decrease with increasing salinities, and that this will result in further degradation of their ecosystem values. Despite the fact that the Salada de Chiprana represents a strongly anthropogenic-influenced salt lake that forms part of a cultural landscape, its specific ecosystem biodiversity and ecosystem properties merit full protection in this global context. Therefore, management actions taken in 1993 allowed preventing direct introduction of surplus irrigation water into the lake. Although these management actions have resulted in beautiful, albeit variable, ecosystem properties from 1995–2006, I think that the current trend of increasing salinities and decreasing lake surface may jeopardize its conservation in the future. **Figure 3** includes a prediction of surface and bathymetry when salinities will reach 120 g TDS L^{-1}, which will imply the demise of the brine shrimps and the disappearance of the *C. chthonoplastes* microbial mats. This is expected to occur before 2020, if no further measures were taken.

The main question is now whether a conservation scheme for the lake La Salada de Chiprana should be based on the conservation of specific communities and ecosystem properties or should priority be given to reducing the human impacts and reestablishment of the natural climatic and hydrological factors? The latter is not really feasible in the context of the human influenced landscape. Therefore, it has been envisioned to add additional clean water from the Civán canal into the lake to compensate for the systematic loss of volume and first experiments have been performed in 2014.

Acknowledgements

The writing and publication of this paper was supported by the EU project ECOPOTENTIAL. ECOPOTENTIAL has received funding from the European Union's Horizon 2020 research and innovation program under Grant Agreement No. 641762. I wish to acknowledge the collaboration with many different colleagues and students with whom I shared many pleasant moments during fieldwork in lake La Salada de Chiprana, among which I would like to mention Maria Carmen Guerrero, Carlos Montes, Alfredo Legaz, Antonio Camacho, Henk Jonkers, Dirk de Beer, Rebecca Ludwig, Ami Bacher, Luisa Falcón, Estelle Masseret, Emilie Le Floc'h, Thierry Bouvier, Cécile Roques, and Pascale Gautret.

Author details

Rutger De Wit

Address all correspondence to: rutger.de-wit@umontpellier.fr

Centre for Marine Biodiversity, Exploitation and Conservation (MARBEC), Centre National de la Recherche Scientifique, Université de Montpellier, IRD and Ifremer, Montpellier, France

References

[1] Williams WD. Environmental threats to salt lakes and the likely status of inland saline ecosystems in 2025. *Environmental Conservation 2002*; 29: 154–167.

[2] Williams WD. The conservation of salt lakes: important aquatic habitats of semi-arid regions. *Aquatic Conservation: Marine and Freshwater Ecosystems* 1993; 3: 71–72.

[3] Grant WD, Tindall BJ. The alkaline saline environment. In: Microbes in Extreme Environments. Academic, London, 1986.

[4] Javor B. Hypersaline Environments, Microbiology and Biogeochemistry. ISBN 3-540-50666-7. Springer-Verlag, Berlin, 1989.

[5] Guerrero MC, De Wit R. Microbial mats in the inland saline lakes of Spain. *Limnetica* 1992; 8: 197–204.

[6] Valero-Garces BL, Navas A, Machin J, Stevenson T, Davis B. Responses of a saline lake ecosystem in a semiarid region to irrigation and climate variability-The history of Salada Chiprana, central Ebro basin, Spain. *Ambio* 2000; 29: 344–350.

[7] Berga A, Cabrera M, Legaz A. Medidas correctoras en el funcionamiento hídrico de la Laguna Salada de Chiprana. *Actas de las "Jornadas sobre el futuro de las Saladas de los Monegros y el Bajo Aragón"*. (English translation: Correcting measures for the hydrody-

namic functioning of La Salada de Chiprana. Proceedings of the workshop dedicated to the future of the salt lakes In Monegros and Bajo Aragón.) Ediciones del Grupo Cultural Caspolino, 1994.

[8] Vidondo B, Martinez B, Montes C, Guerrero MC. Physicochemical characteristics of a permanent Spanish hypersaline lake—La-Salada-De-Chiprana (NE Spain). *Hydrobiologia* 1993; 267: 113–125.

[9] De Wit R, Guerrero MC, Legaz A, Jonkers HM, Blocier L, Gumiaux Ch, Gautret P. Conservation of a permanent hypersaline lake: management options evaluated from decadal variability of *Coleofasciculus chthonoplastes microbial mats. Aquatic Conservation: Marine and Freshwater Ecosystems* 2013; 23: 532–545. DOI: 10.1002/aqc.2319

[10] Guerrero M-C, Balsa J, Pascual M, Martínez B, Montes C. Caracterización limnológica de la laguna Salada de Chiprana (Zaragoza, España) y sus comunidades de bacterias fototróficas. (English translation: Limnological characterizationof Chiprana (Zaragoza, Spain) and its phototrophic communities.) *Limnetica* 1991; 7: 83–96.

[11] Vila X, Guyoneaud R, Cristina XP, Figueras JB, Abella CA. Green sulfur bacteria from hypersaline Chiprana Lake (Monegros, Spain): habitat description and phylogenetic relationship of isolated strains. *Photosynthesis Research* 2002;71: 165–172 doi: 10.1023/A: 1014915900644

[12] Alexander B, Imhoff JF. Communities of green sulfur bacteria in marine and saline habitats analyzed by gene sequences of 16S rRNA and Fenna-Matthews-Olson protein. *International Microbiology* 2006; 9(4): 259–266.

[13] Vanhaecke P, Siddall SE, Sorgeloos P. International study on Artemia. XXXII. Combined effects of temperature and salinity on the survival of Artemia of various geographical origins. *Journal of Experimental Marine Biology and Ecology* 1984; 80: 259–275.

[14] Diaz P, Guerrero MC, Alcorlo P, Baltanás A, Florin M, Montes C. Anthropogenic perturbations to the trophic structure in a permanent hypersaline shallow lake: La Salada de Chiprana (north-eastern Spain). *International Journal Salt Lake Research* 1998; 7: 187–210.

[15] Siegesmund MA, Johansen JR, Karsten U, Friedl T. *Coleofasciculus* gen. nov. (Cyanobacteria): morphological and molecular criteria for revision of the genus Microcoleus Gomont. *Journal of Phycology* 2008; 44: 1572–1585.

[16] Jonkers HM, Ludwig R, De Wit R, Pringault O, Muyzer G, Niemann H, Finke N, De Beer D. Structural and functional analysis of a microbial mat ecosystem from a unique hypersaline inland lake: "La Salada de Chiprana" (N.E. Spain). *FEMS Microbiology Ecology* 2003; 44: 175–189.

[17] Buffan-Dubau E, De Wit R, Castel J. Feeding selectivity of the harpacticoid copepod *Canuella perplexa* in benthic muddy environments demonstrated by HPLC analyses of chlorin and carotenoid pigments. *Marine Ecology Progress Series* 1996; 137: 71–82.

[18] Bachar A, Omoregie E, De Wit R, Jonkers HM. Diversity and function of *Chloroflexus*-like bacteria in a hypersaline microbial mat: phylogenetic characterization and impact on aerobic respiration. *Applied and Environmental Microbiology* 2007; 73: 3975–3983.

[19] De Wit R, Gautret P, Bettarel Y, Roques C, Marlière C, Ramonda R, Nguyen Thanh T, Tran Quang H, Bouvier T. Viruses occur incorporated in biogenic high-Mg calcite from hypersaline microbial mats. *PLoS One* 2015. doi: 10.1371/journal.pone.0130552

[20] Camacho A, De Wit R. Effect of nitrogen and phosphorus additions on a benthic microbial mat from a hypersaline lake. *Aquatic Microbial Ecology* 2003; 32: 261–273.

[21] De Wit R, Falcón L, Charpy-Roubaud C. Heterotrophic dinitrogen fixation (acetylene reduction) in phosphate-fertilised *Microcoleus chthonoplastes* microbial mat from the hypersaline inland lake "la Salada de Chiprana" (NE Spain). *Hydrobiologia* 2005; 534: 245–253.

[22] Bolhuis H, Severin I, Confurius-Guns V, Wollenzien UI, Stal LJ. Horizontal transfer of the nitrogen fixation gene cluster in the cyanobacterium *Microcoleus chthonoplastes*. *ISME Journal* 2010; 4(1): 121–130. doi: 10.1038/ismej.2009.99

[23] Prósper ER. Las Carofitas de España. Singularmente las que crecen en sus estepas. (English translation: The charophytes of Spain, particularly those that grow in its steppes.) Madrid, 1910.

[24] Cirujano S, García Murillo P, Meco A, Fernández Zamudio R. Los carófitos ibéricos. (English translation: The Iberican charophytes). *Anales Jardín Botánico Madrid* 2007; 64: 87–102.

[25] Scheffer M, Carpenter S, Foley JA, Folke C, Walker B. Catastrophic shifts in ecosystems. *Nature* 2001; 413, 591–596 doi:10.1038/35098000

[26] Ludwig R, Al-Horani F, De Beer D, Jonkers HM. Photosynthesis-controlled calcification in a hypersaline microbial mat. *Limnology and Oceanography* 2005; 50: 1836–1843.

Section 2

Lake Biology and Health

Chapter 4

Shallow Lakes of the Mexican Central Plateau: Assessing their Health Condition with Oxidative Stress Biomarkers in Sentinel Organisms

Eugenia López-López, Jacinto Elías Sedeño-Díaz and Ricardo A. Ruiz-Picos

Additional information is available at the end of the chapter

http://dx.doi.org/10.5772/63350

Abstract

In the Mexican Central Plateau (MCP), due to their long history of geologic instability, numerous fluvial systems that were blocked formed extensive shallow lakes. Environmental conditions of this area have favored the agricultural land use and the settlement of great industrial corridors and cities. Human activities in MCP are largely sustained by intense water use that has led to a high deterioration in the water bodies of this area. We analyze the water quality of two selected shallow lakes of the MCP: Yuriria Lake and Xochimilco Lake and early warning biomarkers of native sentinel species of each lake. Both studied lakes are influenced by the input of complex mixtures of pollutants. We assess water quality index and a set of oxidative stress biomarkers in native endemic species of each lake. Results showed that the input of xenobiotics and changes in the periods of dry and rains in the shallow lakes studied provoke a stronger response in sentinel organisms because dilution effects are minimal in a small water column. Furthermore, resuspension of sediments in shallow lakes can release pollutants to the water column that could exert damage to the health condition of the aquatic biota compromising the survival of endemic sensitive species.

Keywords: Yuriria Lake, Xochimilco Lake, lipid peroxidation, antioxidant defenses, water quality, *Ambystoma mexicanum*, *Chirostoma jordani*

1. Introduction

The Mexican highland has capture the interest of several researchers as this area is considered a biodiversity hot spot [1] and, also, is a biogeographic transition zone between the Nearctic and Neotropical zones. Furthermore, this area has a recent volcanic origin that provides singular habitats giving rise to particular evolutionary process to the development of endemic species [2].

The Mexican Plateau is a large arid-to-semiarid plateau that occupies the northern and central Mexico, ranges from the United States border in the north to the Trans-Mexican Volcanic Belt in the south, and is bounded by the Sierra Madre Occidental and Sierra Madre Oriental to the west and east, respectively. A low east-west mountain range in the state of Zacatecas divides the plateau into northern and southern sections: the Northern Plateau and the Central Plateau.

The Central Plateau (Southern Plateau and also called "Mesa de Anáhuac") is a high land lying at elevations from 1800 to 2300 m, which contains numerous valleys originally formed by ancient lakes [3]. This area was dominated by high volcanic activity; consequently, numerous fluvial systems blocked have formed extensive shallow lakes and swamps on the plateau [4]. The ancient name of this area in Nahuatl language was "Anáhuac," which means "Land on the Edge of the Water." These lakes have faced interesting evolutionary process that leads to the development of endemic species [5].

In this zone, a temperate climate, relatively abundant rainfall, and rich alluvial and volcanic soils create favorable agricultural conditions, and much of the land supports extensive farming as well as cattle grazing in some of the drier basins. Land uses are characterized by agriculture (corn, sorghum, beans, wheat, and sugarcane), industry (sugar refineries and grain mills, mining, textile, oil, chemicals, among others), and urban (the area is densely populated and encompasses major urban centers of Mexico including Mexico City) [6, 7]. This area harbors more than 11% of the total population of Mexico, approximately 11 million people, in <3% of Mexican territory, making this region one of the most densely populated in the country. The human activities and development in Central Mexico are largely sustained by intense water use that has led to a high deterioration in the water bodies of this area [8]. Although the most distinctive feature of the Mexican hydrology involves the presence of fluvial systems, the Central Plateau offers the opportunity to study the scarce but interesting lakes originated along with the Trans-Mexican Volcanic Belt.

The agriculture, industrial, and urban activities provoke the input of anthropogenic pollutants (metals, hydrocarbons, pesticides) to aquatic systems, which form complex mixtures of xenobiotics, with the potential to exert many adverse effects on receiver ecosystems [9]. The assessment of the stress effects on the health of the biota in water bodies is very difficult given that exposure to multiple stressors produces both antagonistic and synergistic effects [10]. The bioindicator approach is a validated bioassessment method that monitors responses of key aquatic organisms (sentinels) as integrators of stress effects and as sensitive response (early warning) indicators of environmental health [11]. This integrated approach involves measuring early warning biomarkers and indicators at different levels of organization, from the individual to the population or community level. When applied in field situations, both of these biological indicators can be extrapolated to enable predictions of damage at higher levels of organization that may not be reversible. Some authors [12] have designed protocols for rapid assessment of the health condition, recommending the use of sentinel species to allow monitoring of the response of a biological marker as well as population-based indicators of various environmental stressors.

Early warning biomarkers are useful tools for environmental assessment, which targets analysis of quantification at different levels of organization: molecular, biochemical, physiological, and morphological changes resulting from stress generated by environmental modifications [13]. Several stressors exert damage via oxidative stress, defined as the imbalance between the production of reactive oxygen species (ROS) and antioxidant defense, causing damage to lipids, proteins, carbohydrates, nucleic acids, as well as to the cell membrane [14].

Various xenobiotics are known to exert oxidative stress due to their redox cycling properties and their potential to produce ROS [14]. Lipid peroxidation (LPO) is one of the most commonly used biomarkers for evaluating oxidative stress; it reflects the action of ROS on lipids and is quantified by the presence of lipid metabolites, such as malondialdehyde. Aerobic organisms have developed antioxidant defenses to eliminate or prevent cell damage caused by ROS [15]; some of these defenses include various enzymes such as superoxide dismutase (SOD), catalase (CAT), and glutathione peroxidase (GPx) [16].

For this research, we study the water quality of two selected shallow lakes of the Mexican Central Plateau (MCP): Yuriria Lake (Lerma Basin) and Xochimilco Lake (basin of México), and the oxidative stress responses (lipid peroxidation and antioxidant enzymes) of native sentinel species of each lake, the silverside fish *Chirostoma jordani* (in Yuriria Lake), and the amphibian (Axolotl) *Ambystoma mexicanum* (in Xochimilco Lake).

2. Description of the studied lakes

Yuriria Lake is located in the Mexican Central Plateau in the Lerma–Chapala basin (**Figure 1A**), at 1740 m.a.s.l., with an area of 66 km^2 and a maximum depth of 3.2 m. It is an artificial lake constructed by a diversion of Lerma river, which is their main tributary, the same that carries pollutants such as agrochemicals, heavy metals, organic matter, waste from mining activities, livestock, industrial, urban and rural wastewaters [17]. The western end of the lake receives water intermittently from the channel La Cinta from Cuitzeo Lake, where there are records of the presence of metals such as arsenic, iron, and zinc, as well as discharges from industrial and domestic waste that inputs to the lake [18]. Also, at the western end, the lake receives input from the channel La Cienega from textile and tanning facilities which increases BOD_5, ammonia and total suspended solids, fats and oils, phenols, sulfur, chromium, and other heavy metals [19]. This lake is included in the list of Ramsar wetlands since 2004.

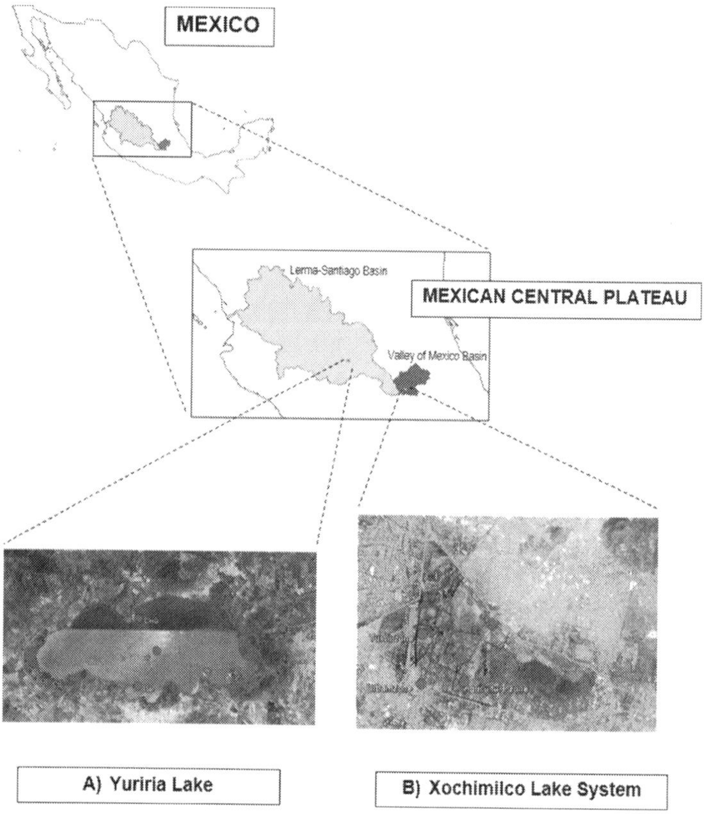

Figure 1. Location of (A) Yuriria Lake and (B) Xochimilco Lake. Circles represent the sampling sites.

The rainy season at Yuriria Lake is from May to September with mean values of 37.5–78.9 mm [20]. In the study period, the lake faced an atypical prolonged drought and the rainfall was delayed with monthly averages of 15 and 16.5 mm in May and August 2009, respectively, and lasted until November (with 10 mm), with values lower than the historical minimum average [21]. Moreover, during May 2010, the mean monthly rainfall was 7 mm; therefore, in this study, the rainy season included the months from May to November 2009 and the dry season from December 2009 to May 2010.

Monitoring was conducted in May, August, and November 2009 and February and May 2010. Seven study sites were established to collect water samples (**Figure 1**): three in the lake (C, CHG, ISL), the three tributaries of the lake (CC, CIE, the diversion channel of the Río Lerma L), and the effluent (E). Specimens of the native fish silverside *Chirostoma jordani* were collected in CHG and ISL with a net (1.5 m high and 4 m in length and 0.8 cm mesh) and a catch effort of 1 h. At each study site, 30 organisms were dissected in the field to get the gills, liver, and muscle for biomarkers assessment. The tissues obtained were transported in liquid nitrogen to the laboratory for later processing.

Xochimilco Lake is a shallow urban water body located in the south of Mexico City with an extension of 26.57 km^2 that belongs to the basin of Mexico. The climate is temperate, ranging from 13 to 25°C, with an average annual temperature of 16°C, an average annual rainfall between 700 and 900 mm, and the rainy season is during the summer (May to October). The lacustrine zone of Xochimilco has a long historical and cultural wealth that includes its high biodiversity and the floating gardens or "Chinampas" [22]. This lake has been one of the main water supplies for Mexico City. Nevertheless, since 1971, the inputs of wastewater in the lake from a treatment facility have provoked water quality depletion and the ecosystem deterioration [23]. Xochimilco Lake is included in the Ramsar list of wetlands since 2004. Three study sites were examined in Xochimilco Lake considering the main land uses that have impact on water quality and generate contrasting conditions: the urban zone (UZ), the tourist zone (TZ), and the agricultural zone (AZ) (**Figure 1B**). Monitoring was conducted in three periods of the year: rainy season (October 2008), the cold drought season (January 2009), and the warm drought season (May 2009). The sentinel organism studied in this lake was the "axolotl" *Ambystoma mexicanum*, a native amphibian that is neotenic and endemic of the basin of México. The species faces a reduction in its populations and is considered critically endangered [24]. The studied organisms were donated specimens that are cultivated in the Unit of Management from the Center of Biological and Aquaculture Researches (CIBAC). Data of oxidative stress biomarkers are from bioassays with the amphibian *A. mexicanum* exposed to elutriates from the three study sites.

3. Acclimatization of *Ambystoma mexicanum*

Acclimatization of organisms: Juvenile organisms of *A. mexicanum* (7 ± 1 cm length and weight of 4.0 ± 0.5 g) were used for the bioassays; organisms were donated by the CIBAC. For acclimatization, organisms were maintained in aquaria for 30 days with hard reconstituted water (100 mg/L $CaCO_3$) [25]. The oxygen saturation levels in the aquaria were maintained by aeration, the temperature was 17 ± 1°C with a photoperiod of 12 h light: 12 h dark. Organisms were fed with *Tubifex* sp. in a daily ration equal to 10% of the wet weight of the amphibians.

4. Sediment elutriation

Sediment elutriates have been useful for evaluating the risk that sediment contamination can exert on aquatic biota and the potential contributions from this matrix to the water column [26]. Elutriates were prepared following the procedure of the USEPA [27]. Water (hard reconstituted water with 100 mg/L $CaCO_3$) was added to sediments in a volume ratio of 1:4 at room temperature (20 ± 2°C). After the correct ratio was achieved, the mixture was stirred vigorously (100 rpm) for 30 min and then was allowed to settle for 1 h. The supernatant was siphoned off without disturbing the settled material and centrifuged to remove particulates prior to chemical analysis (2000 rpm, 30 min). This procedure was done for each study site (UZ, TZ, and AZ) for bioassays.

5. Water Quality Index

In the studied lakes, at each site and study period, dissolved oxygen (mg/L), conductivity (mS/cm), water and air temperature (°C), and pH were recorded *in situ* using a multiparameter probe quanta. Duplicate water samples were collected and transported for their laboratory analysis in dark and cool conditions. In order to quantify hardness ($CaCO_3$ mg/L), color (Pt–Co units) and nitrates (NO_3 mg/L) were determined in the laboratory with a spectrophotometer HACH DRL/2500. The biochemical oxygen demand (BOD_5 mg/L), alkalinity (mg/L), chloride (mg/L) and total and fecal coliform (most probable number, MPN) were quantified according to the APHA [25] procedures. Water Quality Index (WQI) was estimated according to [28], which evaluates on a scale of 0–100.

6. Oxidative stress biomarkers

The set of oxidative stress biomarkers in the sentinel organism of each studied lake includes the level of lipid peroxidation (LPO), and the antioxidant enzymes superoxide dismutase (SOD), catalase (CAT), and glutathione peroxidase (GPx). LPO was determined using the method of [29] in homogenates of the study tissues, previously washed with Tris–HCl, pH 7.4. The results were determined by reading the absorbance of the pink adduct obtained at 535 nm using the molar extinction coefficient of 1.56×10^5 M^{-1} cm^{-1}. CAT was determined using the technique of [30], measuring the consumption of H_2O_2 at 480 nm. SOD activity was evaluated by reading absorbance at 560 nm by the method of [31]. The enzymatic activity of SOD was expressed as units (U) per mg protein. CAT activity was calculated as the rate constant of the first order of the decomposition of H_2O_2. GPx activity was measured with the method proposed by [32] using cumene hydroperoxide as a substrate, and the oxidation of NADPH was determined spectrophotometrically at 340 nm reading for 4 min. The enzyme activity was calculated as nmol NADPH oxidized per minute per mg protein, using the molar extinction coefficient of 6.22×10^6 M^{-1} cm^{-1}. All assays were assessed by triplicate.

7. Statistical analysis

Analysis of variance (ANOVA) was performed to determine the existence of differences: in biomarker responses between sites and periods of study. All statistical analyzes were performed using the software XLSTAT-Pro 2015.

http://dx.doi.org/10.5772/63350

8. The study case of Yuriria Lake

8.1. Water quality

Mean WQI values ranged from 55 to 70, with a global mean of 65.85 ± 4.74 for all sites and study periods. However, in the limnetic zone, mean values ranged from 63 to 70, and the center of the lake (C) was the site with the highest value. The tributaries, especially CC, showed the lowest average value (55), and the global mean of tributaries was 58.3 ± 3.5. The mean value for the effluent was 61 (**Figure 2A**). Mean values of WQI for periods of study varied from 57 to 66, with an overall mean of 61 ± 3.28; August was the month with the lowest WQI and May 2010 had the highest (**Figure 2B**).

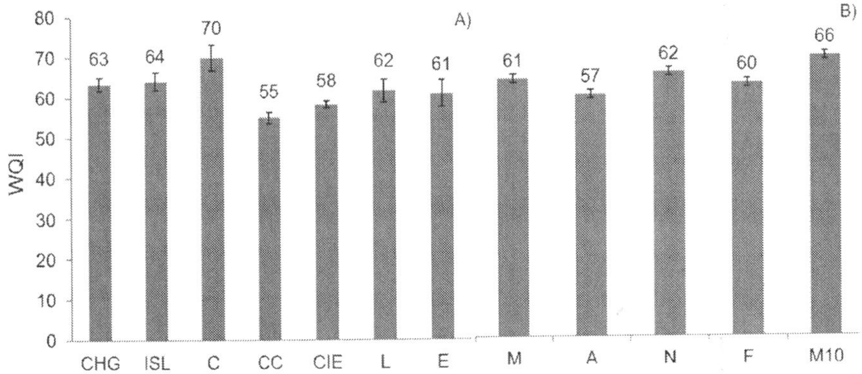

Figure 2. Mean values of WQI in Yuriria Lake. (A) Study sites and (B) periods of study.

8.2. Oxidative stress biomarkers in the fish silverside *Chirostoma jordani*

In general, the gills presented higher values of LPO than liver and muscle. In gill, May 2009 was the month with the highest levels and November the lowest ($p < 0.05$) (**Figure 3A**). In the liver, the level of LPO increased markedly and significantly during November and February in the ISL and May 2010 at both sites, compared to other seasons ($p < 0.05$) (**Figure 3A**). In muscle, the highest level of LPO was observed during February and May 2010 at both sites, while for the rest of the season, LPO values did not differ significantly (**Figure 3A**).

The activity of SOD and GPx significantly increased during November ($p < 0.05$), mainly in the gills, compared to the rest of the seasons (**Figure 3B, C**). CAT activity also showed higher values during November; however, its highest value was found in gills during February in CHG (**Figure 3D**). In general, for both sites, the activity of antioxidant enzymes decreased from February to May 2010 ($p < 0.05$) (**Figure 3B, C**).

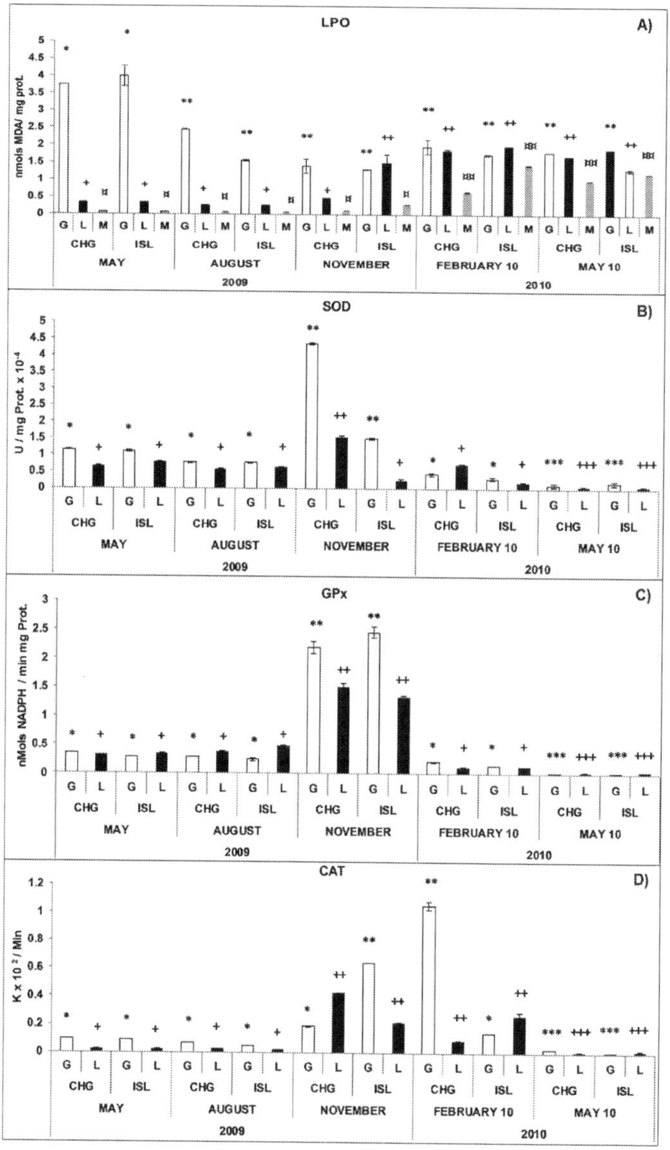

Figure 3. Oxidative stress biomarkers of *C. jordani*. (A) Lipid peroxidation (LPO) in gills (G), liver (L), and muscle (M); (B) superoxide dismutase activity (SOD); (C) glutathione peroxidase activity (GPx); and (D) catalase activity (CAT). * indicates significant differences between sites and seasons, gills (*), liver (+), and muscle (¤). The level of significance was put at $p < 0.05$.

8.3. Somatic indices

The assessment of somatic indices revealed that the condition factor (K) displayed small variation between sites and between periods with no significant differences ($p < 0.05$). The hepatosomatic index (HSI) showed significant differences between sites in August and between periods in May, and values were significantly lower than those of other periods ($p < 0.05$). The gonadosomatic index (GSI) showed the greatest variation, with a clear reproductive peak during May 2010, which differed significantly ($p < 0.05$) from the values found in the other seasons (**Figure 4**).

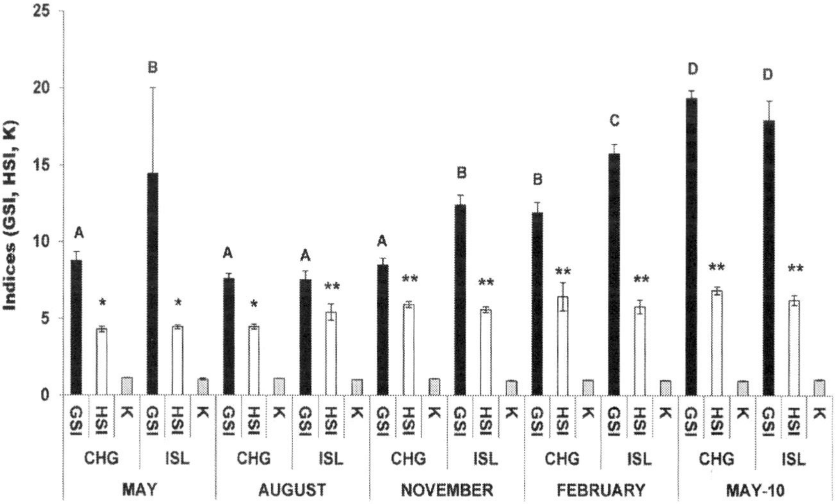

Figure 4. Somatic indices for *C. jordani*. The letters indicate differences statistically significant between GSI, and (*) indicates differences statistically significant between the HSI. The level of significance was put at $p < 0.05$.

9. The study case of Xochimilco Lake

9.1. Water quality

Xochimilco Lake has highly mineralized water and fertilized; conductivity showed higher values (mean value of 791 mS/cm), and the water is highly alkaline (137–288 mg/l), rich in carbonates ($CaCO_3$ 69–120 mg/l) and sulfates (38–67.5 mg/l). The WQI fluctuated from 45.78 to 61.87 in the UZ in June and in the AZ in January, respectively, with a global mean value of 54.48 (**Figure 5A, B**). There were no significant differences between study sites, neither between periods; however, the AZ showed a trend to reach the higher values, particularly in November and January (**Figure 5C, D**).

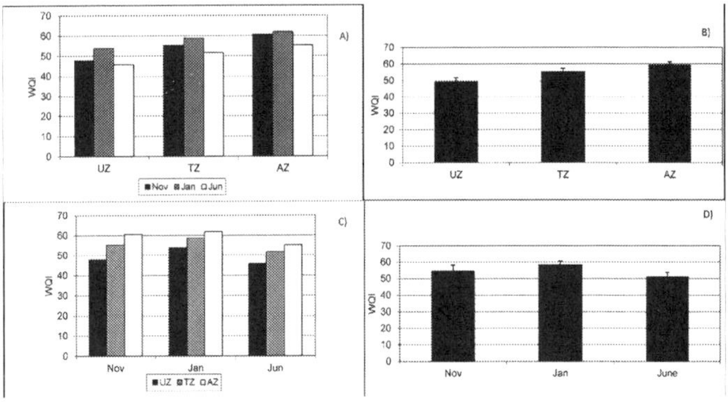

Figure 5. Mean values of WQI in Xochimilco Lake. (A) Study sites, (B) mean values by study sites, (C) periods of study, and (D) mean values by periods of study.

9.2. Oxidative stress biomarkers in the axolotl *Ambystoma mexicanum*

The LPO levels in gills showed the higher values in June, at the beginning of the rainy season, in the three study sites, while LPO levels in liver showed the higher values in AZ in November, in the UZ and TZ in January, and in TZ in June (**Figure 6A**). On the other hand, the antioxidant activities of SOD and GPx showed a depletion in January (during the winter) and June (rainy season) and an increase in activity in November, in all tissues and study sites (**Figure 6B, C**). However, CAT activities in gills showed a very slight fluctuation between periods and study sites, but CAT activity in liver showed a trend to increase in January (in TZ and AZ) and in June in TZ (**Figure 6C**).

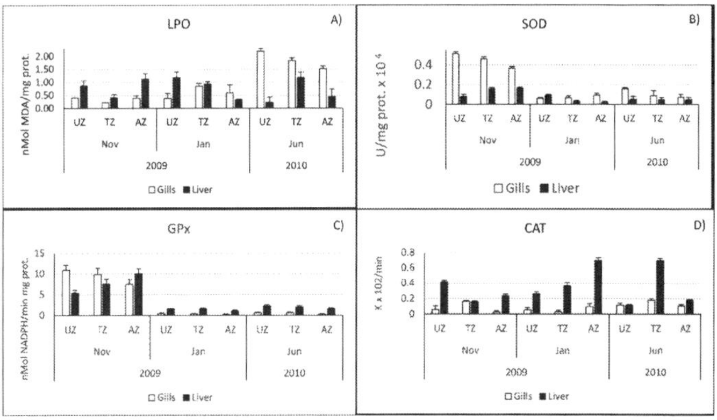

Figure 6. Oxidative stress biomarkers of *A. mexicanum*. (A) Lipid peroxidation (LPO) in gills (G) and liver (L), (B) superoxide dismutase activity (SOD), (C) glutathione peroxidase activity (GPx), and (D) catalase activity (CAT).

10. The importance of using biomarkers in sentinel organisms to assess environmental lake conditions

Yuriria Lake displays the general problem of water quality in the Mexican Central Plateau and the Lerma–Chapala basin, urban, and industrial wastewater discharges, and leachates of agrochemicals [33]. Particularly, the middle Lerma (where Yuriria Lake is located) is recognized as the most polluted area, with WQI scores between 41.1 and 54.2 in 1999 [8, 17]. In the Lake, the WQI ranged from 55 to 70, showing higher values than those of the Lerma River. All the tributaries showed lower WQI, and previous studies recorded that the type of pollution that inputs to the lake differs, forming a mix of xenobiotics with potential to exert damage to the aquatic biota. The tributaries CC and CIE receive discharges from a textile industry, which produces various alkaline residues [34]. The tributary L, which showed the lowest WQI scores, receives discharges from different pollution sources, including oil, mining, and waste disposal from various industries [18].

The toxic effects of the mixture of xenobiotics in aquatic ecosystems depend on the bioavailability and persistence of contaminants; the susceptibility of organisms to accumulate them; and the interference of these compounds with biochemical, physiological, and ecological processes [14]. In Yuriria Lake, our results suggest that the silverside *C. jordani* faces oxidative stress, indicating that the xenobiotics in the lake are able to produce pro-oxidants. The highest LPO levels were observed in gills and liver. Gills are the first organ of contact with water and thus are exposed directly to any xenobiotic in the aquatic environment.

Exposure to various xenobiotics can promote the formation of ROS and induce oxidative stress [16]. An increase in the level of LPO in liver was observed at the end of the rainy season (November) and during the dry season (February and May 2010), making evident the seasons when the pro-oxidant agents in Yuriria Lake trigged oxidative stress.

An increase in LPO levels in fish can trigger an antioxidant response as a defense mechanism to prevent cell damage caused by pro-oxidant agents [15] and could be expressed as increased or depleted in antioxidant enzyme activities [13]. In both cases, the result is impairment to the antioxidant system. In *C. jordani* inhabiting Yuriria Lake, both responses were detected: stimulation of the antioxidant activity in November, when LPO levels in liver and muscle began to increase; and depletion in antioxidant activity during May 2009 with the highest levels of LPO in the gills, and May 2010 when LPO in liver and muscle increased. Some authors [35] have indicated that complex mixtures of contaminants entering water bodies are not stable in time and space. Others [36] have found that a catfish exposure to metals such as isolated Cu, Zn, Pb, Cd, and As can provoke an increase in GPx activity and inhibition of CAT, but in combination, such as Cu and Zn, CAT activity increases. In addition, an increase in CAT activity has been found in fish exposed to polychlorinated biphenyls and polycyclic aromatic hydrocarbons and herbicides such as paraquat and glyphosate [37]. In our study, in November, the highest GPx and SOD activities were detected, but in May, these enzymes showed a decrease; these responses show that the fish is able to adapt to the fluctuating stressors in the lake by inducing additional synthesis of antioxidant enzymes to regulate oxidative stress.

The coupled biological responses of *C. jordani* with environmental fluctuations allowed the identification of critical periods in the health of *C. jordani*, which occurred in February and May 2010 (dry season), when the higher LPO, the lower the antioxidant response, and the lowest K was observed in contrast to the higher WQI scores detected in the same period. The WQI did not include xenobiotic assessment, and consequently, the biological responses assessed in *C. jordani* (biomarkers, somatic indices) are more sensitive as indicators of the lake condition. The drought has been recognized as one of the critical periods in fish health because during this period, the dilution capacity of aquatic ecosystems is low, which increases the risk of exposure to high concentrations of pollutants [38].

Xochimilco Lake is an urban lake that receives the inputs of pollutants from various sources. Urban lakes are impacted by atmospheric deposition, runoffs from urban areas, and rainfall containing a wide range of pollutants [39]. The present study highlights the damages exerted by sediment elutriates in the amphibian *A. mexicanum*, even when all the components of the mixture of pollutants in elutriates were not identified. The LPO reaction mechanisms can damage the lipids found in the cell membrane, altering their cohesion, flowability, permeability, and metabolic function and lead to membrane instability and cell death [14].

Various studies have reported that LPO is an indicator of the toxic action of contaminants, which leads to loss of cell function by oxidative stress. The LPO results in this study showed it is a suitable biomarker that provides important qualitative and quantitative information about the effects of toxic exposure from elutriated sediments from Lake Xochimilco. The most severe damage were detected in June (the rainy season), in the three study sites. During the rainy season in urban lakes, the runoff can lead to the input of several potential pro-oxidant contaminants that can be incorporated to the sediment and exert LPO. The depletion of the antioxidant enzymes SOD and GPx makes evident that January and June are periods were pro-oxidants are more bioavailable and that can exert a severe damage in the defense mechanisms of the sentinel species. Some authors [40] have found that exposure of amphibians to pollutants increases the production of ROS, causing subsequent alterations in the antioxidant defenses. Amphibians have distinctive characteristics, such as their aquatic–terrestrial life and the semipermeability of their skin, that make these species useful as bioindicators [41]; however, there are few studies evaluating amphibian ecotoxicology from urban ecosystems, as the case of Xochimilco Lake.

11. Concluding remarks

The data presented in the study of Yuriria Lake make evident that due to the shallowness of the lake the changes in the periods of dry and rains exert strong influence in the response of sentinel organisms detected by their biomarkers and bioindicators measurements. The input of xenobiotics in Yuriria lake, being a shallow water body, provokes a stronger response in sentinel organisms because dilution effects are minimal in a small water column; consequently, the health condition of the aquatic biota of this lake is more vulnerable to be affected during long dry periods or by the delay of the rainy season; furthermore, once the rainy season began, the lixiviation of xenobiotics from the basin can be incorporated in the lake and also can provoke stressful conditions, also by the scarce dilution potential of a shallow lake.

In the case of Xochimilco Lake, the results showed that the elutriated sediments from the three study sites of Lake Xochimilco were able to induce damages by LPO and in the antioxidant defense of the amphibian, which can compromise several physiological functions and ultimately the life cycle of the amphibian. Since the runoff of pollutants is a common event in urban lakes and sediment resuspension is a frequent process in shallow lakes, the biological responses observed in *A. mexicanum* after exposure to sediment elutriates reflect the risks that this amphibian faces in Lake Xochimilco, due to the toxicological potential of the sediments.

The integrated assessment of aquatic ecosystem approach (assessing environmental condition and biological responses) is more robust than just the physicochemical approach since offers information about the health condition of the aquatic biota. The studied lakes of the Mexican Central Plateau, a zone with high degree of endemisms of aquatic species, are facing stressful conditions that endangered the health of endemic species; thus, it must be a priority to provide measures of conservation to the improvement of the health condition of these aquatic ecosystems.

Acknowledgements

This study was financed by the Research and Postgraduate Studies Secretariat (SIP) of the National Polytechnic Institute (IPN-México) (Project 20100382) and Federal District government, México (ICyTDF/325/11).

Author details

Eugenia López-López[1*], Jacinto Elías Sedeño-Díaz[2] and Ricardo A. Ruiz-Picos[1]

*Address all correspondence to: eulopez@ipn.mx and eugenia_lopez@hotmail.com

1 National Polytechnic Institute, National School of Biological Science, Laboratory of Aquatic Ecosystem Health Assessment, Laboratorio de Evaluación de la Salud de Ecosistemas Acuáticos, Departamento de Zoología, ENCB, Instituto Politécnico Nacional, Ciudad de México, México

2 Polytechnical Coordination for Sustainability, National Polytechnic Institute, Coordinación Politécnica para la Sustentabilidad, Instituto Politécnico Nacional, Ciudad de México, México

References

[1] Myers N, Mittermeier RA, Mittermeier CG, da Fonseca, GAB, Kent J. Biodiversity hotspots for conservation priorities. Nature. 2000;403:853–858. doi:10.1038/35002501

[2] Ferrari L, Orozco-Esquivel T, Manea V, Manea M. The dynamic history of the Trans-Mexican Volcanic Belt and the Mexico subduction zone. Tectonophysics. 2012;522–523:122–149. doi:10.1016/j.tecto.2011.09.018

[3] Ferrusquía-Villafranca I. Geology of Mexico: a synopsis. In: Ramamoorthy TP, Bye R, Lot A, editors. Biological diversity of Mexico: origins and distribution. New York: Oxford University Press; 1993. pp. 3–107. ISBN: 0-19-506674-X

[4] Ferrari L, Pasquarè G, Venegas-Salgado S, Romero-Rios F. Geology of the western Mexican Volcanic Belt and adjacent Sierra Madre Occidental and Jalisco block. In: Delgado-Granados H, Aguirre-Díaz G, Stock JM, editors. Cenozoic Tectonics and Volcanism of Mexico: Boulder, Colorado. Geological Society of America Special Paper 334; 1999. pp. 65–83. doi:10.1130/0-8137-2334-5.65

[5] Moncayo-Estrada R, Israde-Alcántara I, Garduño-Monroy VH. La cherehuita Hubbsina Turneri De Buen (1941) (Pisces: Goodeidae). Origen, distribución y uso de la regionalización de la cuenca del Lerma. Hidrobiológica. 2001;11(1):1–18. ISSN: 0188-8897

[6] Fregoso DA. Vegetación y uso del suelo. In: Secretaría de Medio ambiente y Recursos naturales, editors. Atlas de la Cuenca Lerma-Chapala. Construyendo una visión conjunta. Instituto Nacional de Ecología; 2006. pp. 118–120. ISBN: 968-817-783-0

[7] García-Romero A, Aceves-Quesada JF, Arredondo-León C. Landform instability and land-use dynamics in tropical high mountains, Central Mexico. Journal of Mountain Science. 2012;9:414–430. doi:10.1007/s11629-009-2307-5

[8] Sedeño-Díaz JE, López-López E. Water quality in the Río Lerma, Mexico: an overview of the last quarter of the twentieth century. Water Resource Management. 2005;21:1797–1812. doi:10.1007/s11269-006-9128-x

[9] Viarengo A, Lowe D, Bolognesi C, Fabbri E, Koehler, A. The use of biomarkers in biomonitoring: a 2-tier approach assessing the level of pollutant-induced stress syndrome in sentinel organisms. Comparative Biochemistry and Physiology. 2007;146:281–300. doi:10.1016/j.cbpc.2007.04.011

[10] Segner H, Schmitt-Janses M, Sabater S. Assessing the impact of multiple stressors on aquatic biota: the receptor's side matters. Environmental Science and Technology. 2014;48:7690–7696. doi:10.1021/es405082t

[11] Adams MS, Greeley MS, Ryon MG. Evaluating effects of contaminants on fish health at multiple levels of biological organization: extrapolating from lower to higher levels. Human and Ecological Risk Assessment. 2000;6:15–27. doi:10.1080/10807030091124428

[12] Galloway ST, Brown JR, Browne AM, Dissanayake A, Lowe D, Jones BM. Ecosystem management bioindicators: the ECOMAN project—a multi-biomarker approach to ecosystem management. Marine Environmental Research. 2004;58:233–237. doi: 10.1016/j.marenvres.2004.03.064

[13] Sherry JP. The role of biomarkers in the health assessment of aquatic ecosystems. Aquatic Ecosystem Health and Management. 2003;6(4):440–443. doi:10.1080/714044172

[14] Van der Oost R, Beyer J, Vermeulen NPE. Fish bioaccumulation and biomarkers in environmental risk. Environmental Toxicology and Pharmacology. 2003;13:57–149. doi:10.1016/S1382-6689(02)00126-6

[15] Valavanidis A, Vlahogianni T, Dassenakis M, Scoullos M. Molecular biomarkers of oxidative stress in aquatic organisms in relation to toxic environmental pollutants. Ecotoxicology and Environmental Safety. 2006;64:178–189. doi:10.1016/j.ecoenv. 2005.03.013

[16] Lushchak V. Environmentally induced oxidative stress in aquatic animals. Aquatic Toxicology. 2011;101:13–30. doi:10.1016/j.aquatox.2010.10.006

[17] López-Hernández M, Ramos-Espinosa MG, Carranza-Fraser J. Análisis multimétrico para evaluar contaminación en el Río Lerma y Lago de Chapala, México. Hidrobiológica. 2007;17:17–30. ISSN: 0188-8897

[18] Villalobos-Castañeda B, Alfaro-Cuevas R, Cortés-Martínez R, Martínez-Miranda V, Márquez-Benavides L. Distribution and partitioning of iron, zinc, and arsenic in surface sediments in the Grande River Mouth to Cuitzeo Lake, Mexico. Environmental Monitoring and Assessment. 2010;166:331–346. doi:10.1007/s10661-009-1005-7

[19] Moeller-Chávez G, Mijaylova-Nacheva P, Escalante-Estrada V. (2002) Evaluación de alternativas para reuso del agua en tres giros industriales. In: Federación Mexicana de Ingeniería Sanitaria y Ciencias Ambientales, AIDIS. México, D.F. México: Gestión inteligente de los recursos naturales: desarrollo y salud; 2002. pp. 1–7.

[20] Secretaría de Agricultura, Ganadería, Desarrollo rural, Pesca y Alimentación (SAGARPA). Carta Nacional Pesquera. [Internet] 2010. Available from: http://www.conapesca.sagarpa.gob.mx/wb/cona/actualizacion_de_la_carta_nacional_pesquera_2010. [Accessed: 2016-02-02]

[21] Servicio Meteorológico Nacional (SMN). Precipitaciones media mensual y acumulada 2009 y 2010. [Internet] 2009. Available from: http://smn.cna.gob.mx/index.php?option=com_content&view=article&id=12&Itemid=77. [Accessed: 2016-02-02]

[22] Ezcurra E. De las chinampas a la megalópolis, México: Fondo de Cultura Económica; 1990. pp. 1–85. ISBN: 9789681668716

[23] Bojórquez CL, Amaro MEJ. Multiple characterization of the water quality of Xochimilco channels. In: Stephan-Otto E. editors. The water in the basin of Mexico. Mexico:

Ecological Xochimilco Park Foundation, Metropolitan Autonomous University; 2003. pp. 281–302.

[24] The IUCN Red List of Threatened Species. [Internet]. 2015. Available from: www.iucn-redlist.org. [Accessed: 2016-03-09]

[25] APHA, AWWA, WEF. Standard methods for the examination of water and waste-water, 21st ed. Washington; APHA; 2005. ISBN: 0-87553-047-8

[26] Abrantes N, Pereira R, de Figueiredo DR, Marques CR, Pereira MJ, Goncalves F. A whole sample toxicity assessment to evaluate the sub-lethal toxicity of water and sediment elutriates from a lake exposed to diffuse pollution. Environmental Toxicology. 2009;24(3):259–270. doi:10.1002/tox.20428

[27] USEPA. Bioassay protocols for assessing acute and chronic toxicity at hazardous waste sites. US Environmental Protection Agency. [Internet]. 1998. Available from: http://nepis.epa.gov/. [Accessed: 2016-03-09]

[28] Dinius SH. Design of an index of water quality. Water Resources Bulletin. 1987;23:833–842.

[29] Buege JA, Aust SD. Microsomal lipid peroxidation. Methods in Enzymology. 1978;52:302–310.

[30] Cohen G, Dembiec D, Marcus J. Measurement of catalase activity in tissue extracts. Analytical Biochemistry. 1970;34:30–38.

[31] Sun Y, Oberley LW, Li Y. A simple method for clinical assay of Superoxide dismutase. Clinical Chemistry. 1988;34(3):497–500.

[32] Lawrence RA, Burk RF. Glutathione peroxidase activity in selenium-deficient rat liver. Biochemical and Biophysical Research Communications. 1976;71(4):952–958. doi: 10.1016/0006-291X(76)90747-6

[33] Jiménez-Cisneros BE. Sustentabilidad, un debate a fondo: Información y calidad del agua en México. Trayectorias. 2007;9:45–56. ISSN: 2007-1205

[34] Centro Panamericano de Ingeniería Sanitaria y Ciencias del Ambiente (CEPIS). Informe Técnico sobre Minimización de Residuos en la Industria Textil. [Internet] 1994. Available from: http://http://www.bvsde.paho.org/bvsacd/scan/029704.pdf. [Accessed: 2016-03-09]

[35] De la Torre FR, Salibian A, Ferrari L. Assessment of the pollution impact on biomarkers of effect of a freshwater fish. Chemosphere. 2007;68:1582–1590. doi:10.1016/j.chemosphere.2007.02.033

[36] Farombi EO, Adelowo OA, Ajimoko YR. Biomarkers of oxidative stress and heavy metal levels as indicators of environmental pollution in African cat fish (*Clarias gariepinus*) from Nigeria Ogun River. International Journal of Environmental Research and Public Health. 2007;4:158–165. doi:10.3390/ijerph2007040011

[37] Mañas F, Peralta L, Raviolo J, García-Ovando H, Weyers A, Ugnia L, González-Cid M, Larripa I, Gorla N. Genotoxicity of glyphosate assessed by the comet assay and cytogenetic tests. Environmental Toxicology and Pharmacology. 2009;28:37–41. doi: 10.1016/j.etap.2009.02.001

[38] Ficke A, Myrick C Hansen J. Potential impacts of global climate change on freshwater fisheries. Reviews in Fish Biology and Fisheries. 2007;17:581–613. doi:10.1007/s11160-007-9059-5

[39] Baek YW, An YJ. Assessment of toxic heavy metals in urban lake sediments as related to urban stressor and bioavailability. Environmental Monitoring and Assessment. 2010;171:529–537. doi:10.1007/s10661-009-1297-7

[40] Zocche JJ, da Silva LA, Damiani AP, Mendoça RA, Peres PB, dos Santos CE, Debastiani R, Dias JF, de Andrade VM, Pinho RA. Heavy-metal content and oxidative damage in *Hypsiboas faber*: the impact of coal-mining pollutants on amphibians. Archives of Environmental Contamination Toxicology. 2014;66(1):69–77. doi:10.1007/s00244-013-9949-6

[41] Venturino A, Rosenbaum E, Caballero de Castro A, Anguiano OL, Gauna L, Fonovich de Schoroeder T, Pechen de D' Angelo AM. Biomarkers of effect in toads and frogs. Biomarkers. 2003;8(3–4):167–186. doi:10.1080/1354700031000120116

Chapter 5

Variations in the Zooplankton Species Structure of Eutrophic Lakes in Turkey

Meral Apaydin Yağci

Additional information is available at the end of the chapter

http://dx.doi.org/10.5772/63749

Abstract

In freshwater ecosystems, zooplanktons provide an important food source for larval fish. Some species of zooplanktons are usually considered as useful indicators of water quality and trophic state, for example, *Filinia longiseta, Keratella cochlearis, Keratella quadrata, Brachionus angularis, etc.* The purpose of this study is to describe the recent zooplankton composition and monthly variations, to compare results with earlier studies, to determine the trophic level of the eutrophic lake and to show eutrophic lakes in Turkey.

Keywords: zooplankton (*Rotifera, Cladocera, Copepoda*), eutrophic, lake, Karamık, Turkey

1. Introduction

Zooplanktonic organisms form the second ring of the food chain in lake ecosystem after phytoplanktonic organism. Zooplanktonic organisms are one of the crucial study subjects in freshwater ecosystems. As forming the basic food chain between primary producers and other higher forms in the founded food chain, zooplankton has a great significance in aquatic ecosystems. Naturally, undergoing a change in number or variety of these organisms affects the living groups on the top of the food chain pyramid [1]. Zooplankton is significant for the nutrition of fish, fry, some invertebrates and birds and in the biodiversity structure of the ecosystem [2–5]. Besides, it is used as an indicator of trophic situation and the quality of water in the lakes [6–10]. Zooplanktonic organisms, which are significant for inland fishery, with some species, are evaluated as indicators of contamination with their sensitivity to environmental changes and forming the great part of the aquatic organisms [11]. Zooplankton species are showed as bioindicator for determining the environmental contamination occurred due to environmental parameters [12, 13]. Among zooplanktonic organisms, *Rotifera* is the primary vertebrate group of freshwater. While *Rotifera* has an important role in many freshwater, *Cladocera* and *Copepoda* group are utilized in culture studies [14]. In eutrophication, distribution and intensity of especially *Copepoda* and *Rotifera* group of organisms among zooplanktonic organisms have been pointed out to be effective [15, 16]. Among zooplanktonic organisms, particularly small-sized *Cladocera,* the dominance of *Rotifera* species means eutrophic lakes; on the other hand, the dominance of *Copepoda* species points out oligotrophic lakes [17, 18]. In addition to this, *Cyclopoida* and *Harpacticoid Copepoda* groups have been used to determine the trophic situation [19, 20]. In the lakes where some dominant zooplankton species are located, the biodiversity of eutrophic lake has been observed as low. Also, zoobentos and *Rotifera* diversity are very low in the areas where eutrophication has been shown very high [21, 22]. *Copepoda,* which is intensely found in freshwater ecosystems, forms the main constituent of planktonic community. *Copepoda* group of organisms have been shown as the indicator species for determining the quality of water [23].

In the sediments of the lakes, the records of the change of the biological communities can be followed constantly, because while protecting the quality and the structure in aquatic ecosystems, it is benefited from the earlier formations. In fact, diatoms have been used in the content of functional implementations to evaluate the paleolimnological past of the trophic structure of the lakes. Besides, ecological communities in freshwater ecosystems have been primarily examined in taxonomically and morphologically. Lately, functional organizations, distribution of the species and different ecological roles have been stated as important features. *Cladocera* is important on the water quality and ecology of the lakes when appeared in the middle level of trophic degree [24]. Also, *Cladocera* has been defined as a good indicator of hydrological dynamism of the water floor and the heterogeneity of the habitat [25]. *Rotifera, Cladocera,* and *Copepoda* are the basic groups of zooplankton in freshwaters. Many *Rotifera* has great effect on the metabolism and rapid turnover of the lacustrine food chain [26, 27]. *Cladocera* group has viability feature in extreme conditions of freshwater [20]. It was observed that the contribution of Rotifer species in zooplankton community might raise the eutrophication [28]. Generally, zooplankton gives immediate reaction to climate change, but endangered populations can be too much insistent on main distribution areas for a long time [29]. Among the zooplankton community, Rotifers, having a high sensitiveness on the conditions of the environment where they live, have a particular importance being an indicator on determining the water quality, contamination, and eutrophication level of the water that contains some species [30]. To fulfill the food need of the people, freshwater fishery is gaining emphasis gradually because of the rising population of the world. Thus, the studies have been speeded up on zooplanktonic organisms, which are one of the main food sources of many fish species in larval stage and forms the basic chain of the food chain in transforming the phytonutrients to proteins in aquatic environment.

2. Eutrophic lakes in Turkey

Eutrophic lakes in Turkey are listed in **Table 1**.

Lake name	Locality	Surface area	Trophic state	References
Beyşehir Lake	Konya, Isparta	650 km²	Mesotrophic-eutrophic	[31, 32]
Eğirdir Lake	Eğirdir, Isparta	468 km²	Mesotrophic-eutrophic	[33]
Marmara Lake	Salihli, Manisa	45 km²	Eutrophic	[34]
Akşehir Lake	Konya	353 km²	Mesotrophic-eutrophic	[35, 36]
Karagöl	Yamanlar, İzmir	2 ha	Eutrophic	[37]
Çavuşçu Lake	Ilgın, Konya	1000 ha	Mesotrophic-eutrophic	[38]
Gala Lake	Enez İpsala	5.6 km²	Eutrophic	[39]
Mogan Lake	Gölbaşı, Ankara	561.2 ha	Mesotrophic-eutrophic	[40]
Karataş Lake	Burdur	11.9 km²	Vulnerable to eutrophic	[5, 41]
Gölhisar Lake	Burdur	400 ha	Mesotrophic-eutrophic	[42]
Ladik Lake	Samsun	141.40 km²	Eutrophic	[10]
Yenişehir Lake	Reyhanlı, Hatay	105.340 m²	Eutrophic	[43]
Terkos Lake	İstanbul	25 km²	Eutrophic	[44]
Çernek Lake	Samsun	589 ha	Eutrophic	[45]
Karamık Lake	Çay, Afyonkarahisar	38 km²	Eutrophic	[1, 46, 47]
Liman Lake	Kızılırmak Delta	50.000 ha	Eutrophic	[48]
Gölcük Lake	Ödemiş, İzmir	125 ha	Eutrophic	[49]
Eber Lake	Bolvadin, Afyonkarahisar	12.500 ha	Eutrophic	[50]

Table 1. List of eutrophic lakes in Turkey.

3. Zooplankton indicators of eutrophic lake

Zooplanktonic organisms are studied from many researchers in terms of taxonomy and ecology. Species revealed in freshwater areas are used as an indicator to determine the water quality, contamination and eutrophication condition [51, 52]. *Rotifera* species are widely stated as an indicator of eutrophic water. *Brachionus calyciflorus, Trichotria tetractis,* and *Filinia longiseta* species are used as an indicator of eutrophication [53]. It is believed that *Rotifera* is dominant in eutrophic water. Because parthenogenetic reproduction and population length are very broad, lakes were seen as convenient for eutrophication for the species of *Sattal Keratella* sp, *Brachionus* sp, *F. longiseta,* and *T. tetractis* [54]. *Brachionus calyciflorus, B. angularis, Keratella quadrata, K. cochlearis, Polyarthra dolichoptera, Euchlanis dilatata, Lecane luna, Pompholyx sulcata, Filinia longiseta, Trichocerca species* (*Rotifera*), *Bosmina longirostris, Chydorus sphaericus,* and *Daphnia cucullata* (*Cladocera*) are among the species characteristic of eutrophic water [55]. It was reported by Bozkurt and Akın [56] that species of *Brachionus quadridentatus, Notholca squamula, Lepadella patella, Lecane bulla, Bosmina longirostris, Alona rectangula, Acanthocyclops robustus,* and *Cyclops vicinus* were known as the indicator species of eutrophication and found mainly in hot zones. In Ballybeg, Crans and Morgan Lakes in Ireland *Bosmina longirostris, Daphnia longispina, D.pulex, Leydigia leydigi* and *Oxyurella tenuicaudis* species while showing a positive relationship with trophic condition variables at the same time are formed in high eutrophic areas [24]. Yet, in the study on the composition of zooplankton in Ladik Lake in 2015, the lake was identified as eutrophic for *Brachionus calyciflorus, B. angularis, Keratella quadrata, K. cochlearis, Bosmia longirostris, Cyclops vicinus,* and *Thermocyclops crassus* [10]. *Keratella cochlearis, Keratella qaudrata, Brachionus angularis* species, among the *Rotifera* group of organisms, are shown as the indicator of eutrophication by some researchers [2, 19, 27, 57]. Among the *Rotifera* group, *Brachionus calyciflorus, B. angularis, Keratella cochlearis, Keratella qaudrata, Conochilus dossuarius, Filinia longiseta, Trichocerca capucina, Trichocerca cylindrica* and *Bosmina longirostris, Graptoleberis testudinaria* among Cladocera group are stated as the most important indicators found in eutrophic conditions [19, 58–61].

While *Brachionus angularis* and *Keratella tropica* species form strongly in eutrophicated water [62], *Brachionus calyciflorus, Daphnia* sp. and *Ceriodaphnia* sp form in eutrophication conditions in Beira Lake [8]. Furthermore, *Cladocerans* and *Cyclopoid Copepods* adapted very well to eutrophic water [63]. In Sakarya River (Turkey) zooplankton, many species are shown as the indicator of eutrophication. As typical eutrophic species *Brachionus* spp., *E. dilatata, F. longiseta, K. cochlearis, K. quadrata, P. quadricornis, B. longirostris, C. sphaericus* and *C. vicinus* in Sakarya River [64]; *B. urceolaris, B. calyciflorus, K. quadrata, E. dilatata, B. longirostris, C. sphaericus, C. vicinus* and *E. serrulatus* in Gölcük Lake were found [48]. *Anuraeopsis fissa, Brachionus angularis. Keratella cochlearis f. tecta, Pompholyx sulcata* and *Filinia longiseta* species are important eutrophication indicators in the lakes of Poland [65].

4. Example of eutrophic lake: Lake Karamık (Çay, Afyonkarahisar, Turkey)

On the fishery of Lake Karamık, the size distribution of pike population, spawning period and nutrition were studied by Aksun [66, 67], and then in 2000s, the fishery of Lake Karamık and determining the biological properties of economical fish species were investigated by Çubuk et al [68]. Lake Karamık is located in 20 km southwest of Çay District in Afyon Province. Karakuş Mountains lie in the south part. The altitude is 1067 m, the acreage is 38 km^2 and the average deepness ranges between 2 and 3 m. Most of the lake is covered with plants living in water and on the surface of the water, and thus, the area for fishery is very limited. Lake Karamık is a very shallow lake showing eutrophic properties [1]. Snow and rain water, and Dipsiz, Aykırı and Kocabaş springs are the water sources of the lake and the outlets of the lake are evaporation and the sinkholes, named as Büyük subatan and Küçük subatan, in the south of the lake. The lake drains its water to Lake Hoyran via sinkholes. The wastewater of Afyon-

Çay SEKA paper mill was emptied to the lake in the past. After the year 2004, the mill was closed and the wastewater was prevented.

5. Study area

5.1. Sample collection and analysis

Monthly sampling was realized in Lake Karamık in three stations between the dates of March 2002 and March 2003. The coordinates of first station are 38° 26′ 10.16″ N, 30° 52′ 14.30″E, second station are 38° 27′ 13.33″N, 30° 51′ 20.05″E and third station are 38° 26′ 44.60″ N, 30° 50′ 36.47″E in Çay District in Afyon Province (**Figure 1**). Zooplankton samples were taken out of the lake via a 55 micron length of plankton mesh, and then, the samples were fixed in a %4 formaldehyde solution. Related sources were used in the sorting and identification of the species of zooplanktonic organisms [69–76]. Olympus model search technique and the invert and stereo model microscopes were used in species identification While making preparations for identification of the species, the samples were taken on the glycerine + formalin mixture dribbled on the slide, and then, the samples, covered with cover glass, were fixed with Canada balsam and taken into the collection In the identification process of the *Rotifera* species taking trophy with bleach, in the identification process of *Copepoda* and *Cladocera*, dissection process was realized. 0.10 and 0.60 mm numbered FST (fine science tools) stainless steel injections were

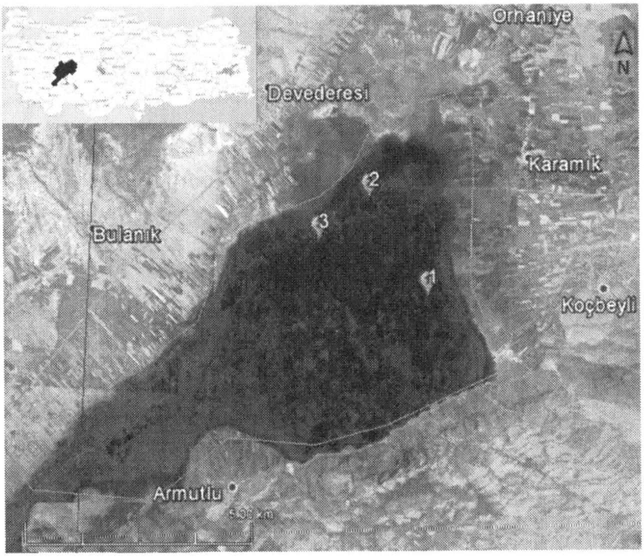

Figure 1. The study area and sampling stations in Lake Karamık.

used in the dissection process. Photographs of the organisms were taken by a microscope connected Nikon brand imaging device.

To determine the trophic index of the lake, *Brachionus*:*Trichocerca* (QB/T) [61] equality was used. According to Sládeček [61], while a quotient between one and two corresponds to mesotrophic conditions and a ratio of >2 is encountered in eutrophic lakes, a quotient of 1 indicates oligotrophic conditions. We used Soyer's [77] frequency index to define the frequency of species in the research area and constant ($F \geq 50\%$), common ($50\% > F \geq 25\%$), or rare ($F < 25\%$) estimated results were taken. $F = m/M \times 100$ is used in this index (F) to evaluate special species that m is the number of stations for the species and M is the number of all stations.

6. Results and discussion

In this study, totally 69 zooplankton species were identified. Thirty-seven of the species belong to *Rotifera* group (%54), 22 to *Cladocera* group (%32) and 10 to *Copepoda* group (%14) (**Figure 2**). The identified taxa were listed according to Ustaoğlu [78] and Ustaoğlu et al. [79].

6.1. Zooplankton composition of Lake Karamık (Çay, Afyonkarahisar, Turkey)

Platyias quadricornis (Ehrenberg, 1832), *Brachionus angularis* Gosse, 1851, *Brachionus quadridentatus* Hermann, 1783, *Brachionus rubens* Ehrenberg, 1838, *Keratella tecta* (Gosse, 1851), *Keratella quadrata* (Müller, 1786), *Notholca acuminata* (Ehrenberg, 1832), *Euchlanis incisa* Carlin, 1939, *Euchlanis meneta* Myers, 1930, *Mytilina mucronata* (Müller, 1773), *Mytilina ventralis* (Ehrenberg, 1830), *Trichotria pocillum* (Müller, 1776), *Trichotria tetractis* (Ehrenberg, 1830), *Macrochaetus collinsii* (Gosse, 1867), *Lepadella ehrenbergi* (Petry, 1850), *Lepadella ovalis* (Müller, 1786), *Lecane elsa* Hauer, 1931, *Lecane luna* (Müller, 1776), *Lecane ohioensis* (Herrick, 1885), (Herrick, 1885), *Lecane bulla* (Gosse, 1886), *Lecane clostrocerca* (Schmarda, 1859), *Lecane cornuta* (Müller, 1786), *Lecane lunaris* (Ehrenberg, 1832), *Lecane curvicornis* (Murray, 1913), *Lecane quadridentata* (Ehrenberg, 1830), *Trichocerca iernis* (Gosse, 1887), *Trichocerca similis* (Wierzeski, 1893), *Ascomorpha saltants* Bartsch, 1870, *Synchaeta pectinata* Ehrenberg, 1832, *Polyarthra dolichoptera* delson, 1925, *Dicranophorus grandis* (Ehrenberg, 1832), *Testudinella patina* (Hermann, 1783), *Conochilus unicornis* Rousselet, 1892, *Hexarthra mira* (Hudson, 1871), *Filinia longiseta* (Ehrenberg, 1834), *Filinia terminalis* (Plate, 1886), *Colletheca* sp, *Daphnia curvirostris* Eylmann, 1887, *Daphnia longispina* O.F. Müller, 1875, *Simocephalus exspinosus* (Koch, 1841), *Simocephalus vetulus* (O.F. Müller, 1776), *Ceriodaphnia quadrangula* (O.F. Müller, 1785), *Ceriodaphnia reticulata* (Jurine, 1820), *Scapholeberis kingi* Sars, 1903, *Bosmina longirostris* (O.F. Müller, 1785), *Eurycercus lamellatus* (O.F. Müller, 1785), *Pleuroxus aduncus* (Jurine, 1820), *Pleuroxus laevis* (Sars, 1862), *Alonella excisa* (Fischer, 1854), *Alonella nana* (Baird, 1850), *Chydorus sphaericus* (O.F. Müller, 1776), *Alona costata* Sars, 1862, *Alona guttata* Sars, 1862, *Alona rectangula* Sars, 1862, *Acroperus harpae* (Baird, 1835), *Graptoleberis testudinaria* (Fischer, 1848), *Biapertura affinis* (Leydig, 1860), *Tretocephala ambiqua* (Lilljeborg, 1900), *Oxyurella tenuicaudis* (Sars, 1862), *Acanthodiaptomus denticornis* (Wierzejski, 1887), *Macrocyclops albidus* (Jurine, 1820), *Eucyclops macruroides* (Lilljeborg, 1901), *Eucyclops macrurus* (G.O. Sars, 1863), *Eucyclops serrulatus* (Fischer, 1851), *Eucyclops speratus*

(Lilljeborg, 1901), *Cyclops strenuus paternonis (Lindberg, 1956), Megacyclops viridis* (Jurine, 1820), *Megacyclops gigas* (Claus 1857), *Nitocra hibernica* (Brady, 1880).

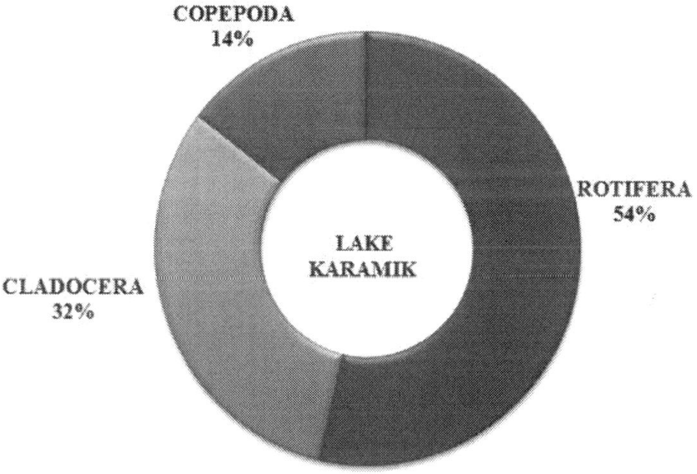

Figure 2. Composition of zooplankton in Lake Karamık.

Among the species identified in the lake, *K. quadrata, L. luna, L. bulla, T. patina, D. longispina, D. curvirostris, E. lamellatus, C. aphaericus, A. rectangula, P. aduncus, A. denticornis, E. macrurus* and *C. strenuus paternonis* were widely found, and *P. quadricornis, B. rubens, K. cochlearis tecta, N. acuminata, E. meneta, M. mucronata, T. tetractis, L. elsa, T. similis, D. grandis* and *A. guttata* were rarely found (**Table 2**). In this lake, 42 species by Gündüz [1], 86 species by Emir and Demirsoy [46], 18 species by Gündüz [80] and 9 species by Kazancı et al. [41] were reported. While 56 species were similarly found as in previous searches [1, 41, 46, 80], 64 species that were determined in previous were not appeared in this study (**Table 3**). First study on zooplankton in the lake started with Gündüz [1] (20 species of *Rotifera* and 22 species of *Cladocera*). Though in the study after 12 years, *Rotifera* group of organisms were widely studied [46] (86 species of *Rotifera*). Eighteen organisms only of *Cladocera* group were listed in the checklist presented in 1997 [80]. However, in the limnologic survey realized in the lake in 1999, a few zooplanktonic organisms were reported (totally nine species). The new species identified in this study (*K. cochlearis tecta, L. ehrenbergi, L. ovalis, L. cornuta, L. curvicornis, T. similis, D. grandis, Colletheca* sp., *B. longirostris, A. guttata, B. affinis, E. macruroides, E. serrulatus, M. gigas)* were indicated with asterisk and bold (**Table 3**). According to frequency index values, four species classified as constant ($F \geq 50$ %), 14 species as common (50 % > $F \geq 25$ %), 51 species as rare ($F < 25$ %). The constant species *K. qaudrata* was identified in all stations with the highest frequency value (74.29%). Of the group *Cladocera, C. sphaericus* (71.43%), *A. rectangula* (60 %), *P. aduncus* (54.29%) were other widespread species (**Table 2**).

Study period	2002																													2003									
Months	March			April			May			June			July			August			September			October			November			December		January			February			March			
Stations/Species	1	2	3	1	2	3	1	2	3	1	2	3	1	2	3	1	2	3	1	2	3	1	2	3	1	2	3	1	2	1	2	3	1	2	3	1	2	3	f %
ROTIFERA																																							
Platyias quadricornis	–	–	–	◇	–	–	–	–	–	–	–	–	–	–	–	–	–	–	–	–	–	–	–	–	–	–	–	–	–	–	–	–	–	–	–	–	–	–	2.86
Brachionus angularis	–	–	–	–	–	–	–	–	–	◇	–	–	❂	◇	❂	◇	–	★	◇	★	❂	–	◇	◇	–	–	–	–	–	–	–	–	–	–	–	–	–	–	31.43
Brachionus quadridentatus	–	–	–	–	–	–	–	–	–	–	–	–	–	–	–	–	–	◇	◇	–	–	–	–	–	–	–	–	–	–	–	–	–	–	–	–	–	–	–	5.71
Brachionus rubens	–	–	–	–	–	–	–	–	◇	–	–	–	–	–	–	–	–	–	–	–	–	–	–	–	–	–	–	–	–	–	–	–	–	–	–	–	–	–	2.86
Keratella tecta	–	–	–	–	–	–	–	◇	–	–	–	–	–	–	–	–	–	–	–	–	–	–	–	–	–	–	–	–	–	–	–	–	–	–	–	–	–	–	2.86
Keratella quadrata	❂	❂	★	❂	❂	–	❂	❂	◇	❂	❂	❂	★	◇	❂	◇	◇	★	★	★	★	◇	★	★	–	–	–	–	◇	–	–	–	–	◇	–	–	–	◇	74.29
Notholca acuminata	–	–	–	–	★	–	–	–	–	–	–	–	–	–	–	–	–	–	–	–	–	–	–	–	–	–	–	–	–	–	–	–	–	–	–	–	–	–	2.86
Euchlanis incisa	–	–	–	–	–	–	–	–	–	–	★	◇	◇	–	◇	–	–	–	◇	–	–	◇	–	–	◇	–	–	–	–	–	–	–	–	–	–	–	–	–	20
Euchlanis meneta																													◇	–	–	–	–	–	–	–	–	–	2.86
Mytilina mucronata	–	–	–	–	–	–	–	–	–	–	–	–	–	–	–	–	–	–	◇	–	–	–	–	–	–	–	–	–	–	–	–	–	–	–	–	–	–	–	2.86
Mytilina ventralis	–	–	–	–	–	–	–	–	–	–	–	–	◇	–	◇	–	–	–	◇	–	–	–	–	–	–	–	–	–	–	–	–	–	–	–	–	–	–	–	8.57
Trichotria pocillum	–	–	–	–	–	–	–	–	–	–	–	–	◇	◇	◇	–	–	–	–	◇	–	–	–	◇	–	–	–	–	–	–	–	–	–	–	–	–	–	–	14.29
Trichotria tetractis	–	–	–	–	–	–	–	–	–	–	–	–	–	–	–	–	–	–	◇	–	–	–	–	–	–	–	–	–	–	–	–	–	–	–	–	–	–	–	2.86
Macrochaetus collinsii	–	–	–	–	–	–	–	–	–	–	–	–	–	–	–	–	–	–	◇	–	–	–	◇	–	–	–	–	–	–	–	–	–	–	–	–	–	–	–	5.71
Lepadella ehrenbergi	–	–	–	–	–	–	–	–	–	–	–	–	–	◇	–	◇	–	–	◇	◇	–	–	◇	◇	–	–	–	–	–	–	–	–	–	–	–	–	–	–	17.14
Lepadella ovalis	–	–	–	–	–	–	–	–	–	◇	–	–	–	–	–	–	–	–	◇	–	–	★	◇	–	–	–	–	–	–	–	–	–	–	–	–	–	–	–	11.43
Lecane elsa	–	–	–	–	–	–	–	–	–	–	–	–	◇	–	–	–	–	–	–	–	–	–	–	–	–	–	–	–	–	–	–	–	–	–	–	–	–	–	2.86
Lecane luna	–	–	–	–	–	–	◇	–	–	◇	–	–	◇	★	★	★	◇	★	◇	★	◇	★	◇	★	–	–	–	–	–	–	–	–	–	–	–	–	–	–	40
Lecane ohioensis	–	–	–	–	–	–	–	–	–	–	–	–	–	◇	◇	★	–	◇	◇	◇	◇	◇	◇	◇	–	–	–	–	–	–	–	–	–	–	–	–	–	–	28.57
Lecane bulla	–	◇	–	–	–	–	–	◇	–	◇	–	–	◇	★	★	◇	–	★	◇	★	★	❂	★	★	–	–	–	–	–	–	–	–	–	–	–	–	–	–	40
Lecane clostrocerca	–	–	–	◇	–	–	–	–	–	★	–	–	◇	◇	–	◇	–	–	◇	◇	–	◇	◇	◇	–	–	★	–	–	–	◇	–	–	–	–	–	–	–	34.26
Lecane cornuta	–	–	–	–	–	–	–	–	–	–	–	–	–	–	◇	–	–	◇	◇	–	◇	–	◇	–	–	–	–	–	–	–	–	–	–	–	–	–	–	–	14.29
Lecane lunaris	–	–	–	–	–	–	–	–	–	–	–	–	–	–	–	◇	–	–	–	◇	–	◇	◇	◇	–	–	★	–	–	–	–	–	–	–	–	–	–	–	17.14
Lecane curvicornis	–	–	–	–	–	–	–	–	–	–	–	–	–	◇	–	–	–	◇	◇	◇	◇	–	◇	–	–	–	–	–	–	–	–	–	–	–	–	–	–	–	17.14
Lecane quadridentata	–	–	–	–	–	–	–	–	–	–	–	–	–	–	–	–	–	–	–	–	★	◇	◇	★	–	–	–	–	–	–	–	–	–	–	–	–	–	–	11.43
Trichocerca iernis	–	–	–	–	–	–	–	–	–	–	–	–	–	–	–	–	–	–	–	–	–	–	◇	◇	–	–	–	–	–	–	–	–	–	–	–	–	–	–	5.71
Trichocerca similis	–	–	–	–	–	–	–	–	–	–	–	–	–	–	–	–	–	–	–	–	–	–	–	◇	–	–	–	–	–	–	–	–	–	–	–	–	–	–	2.86
Ascomorpha saltants	–	–	–	–	–	–	–	–	–	★	–	–	–	–	◇	–	–	◇	◇	–	–	–	◇	◇	–	–	–	–	–	–	◇	–	–	–	–	–	–	–	20
Synchaeta pectinata	★	◇	–	–	◇	–	–	◇	–	–	–	–	–	–	–	–	–	–	◇	–	◇	◇	◇	◇	◇	–	–	–	◇	◇	–	–	–	–	–	–	–	–	34.29
Polyarthra dolichoptera	–	–	–	–	–	–	★	◇	◇	◇	–	–	◇	–	◇	–	–	–	–	–	–	–	–	–	–	–	–	◇	◇	◇	◇	–	–	–	–	–	–	–	28.57

Study period	2002																											2003									
Months	March			April			May			June			July			August			September			October			November			December		January			February			March			
Stations/Species	1	2	3	1	2	3	1	2	3	1	2	3	1	2	3	1	2	3	1	2	3	1	2	3	1	2	3	1	2	1	2	3	1	2	3	1	2	3	f %
Dicranophorus grandis	–	–	–	–	–	–	–	–	–	–	–	–	◇	–	–	–	–	–	–	–	–	–	–	–	–	–	–	–	–	–	–	–	–	–	–	–	–	–	2.86
Testudinella patina	–	–	–	◇	–	–	–	–	–	★	✪	★	★	◇	◇	–	–	◇	◇	★	◇	◇	◇	◇	–	–	–	◇	–	–	–	–	–	◇	–	–	–	–	45.71
Conochilus unicornis	–	–	–	–	–	–	–	–	–	✪	◇	◇	–	◇	–	–	–	–	–	–	–	–	–	–	–	–	–	–	–	–	–	–	–	–	–	–	–	–	11.43
Hexarthra mira	–		–	–	–	–	–	◇	–	–	–	–	◇	◇	–	–	–	–	–	–	–	–	–	–	–	–	–	–	–	–	–	–	–	–	–	–	–	–	8.57
Filinia longiseta	–	–	–	–	–	–	–	–	–	–	–	–	–	◇	◇	–	–	–	–	◇	◇	–	–	–	–	–	–	–	–	◇	–	–	–	–	–	–	–	–	14.29
Filinia terminalis	–	–	◇	–	–	–	◇	–	–	–	–	–	–	–	–	–	–	–	–	–	–	–	–	–	–	–	–	–	–	–	–	–	–	–	–	–	–	–	5.71
Collotheca sp.	◇	◇	–	–	–	–	–	–	–	–	–	–	–	–	–	–	–	–	–	–	–	◇	–	–	–	–	–	–	–	–	◇	◇	–	–	◇	–	–	–	17.14
CLADOCERA																																							
Daphnia curvirostris	–	–	–	–	–	–	–	–	✪	◇	◇	◇	–	–	–	–	–	–	–	–	◇	–	–	–	–	–	–	◇	–	–	✪	★	–	–	–	✪	–	–	25.71
Daphnia longispina	–	–	–	★	–	–	–	◇	–	–	–	–	–	–	–	–	–	–	–	–	–	–	–	–	–	–	–	◇	–	◇	–	–	★	★	✪	–	✪	✪	25.71
Simocephalus exspinosus	◇	–	–	–	–	–	–	–	–	–	–	–	–	–	–	–	–	–	–	–	–	–	–	–	–	–	–	–	–	–	–	–	–	–	–	–	–	–	2.86
Simocephalus vetulus	–	–	–	–	–	–	–	–	–	–	–	–	–	–	–	–	–	–	–	–	–	–	–	–	★	–	–	–	–	–	–	–	–	–	–	◇	◇	◇	11.43
Ceriodaphnia quadrangula	–	–	–	–	–	–	–	–	◇	–	–	–	◇	–	–	–	–	–	–	–	–	–	–	–	–	–	–	◇	–	–	–	–	–	–	–	–	–	–	8.57
Ceriodaphnia reticulata	–	–	–	–	–	–	–	–	–	–	–	–	–	◇	–	–	–	–	–	–	–	–	–	–	–	–	–	–	–	–	◇	–	–	–	–	–	–	–	5.71
Scapholeberis kingi	–	–	–	–	–	–	–	◇	–	◇	–	◇	–	–	–	–	–	–	–	–	–	–	–	–	–	–	–	–	◇	–	–	–	–	–	–	–	–	–	11.43
Bosmina longirostris	–	–	–	–	–	–	–	–	–	–	–	–	–	–	–	–	–	–	–	–	–	–	◇	◇	–	–	–	–	–	–	–	–	–	–	–	–	–	–	5.71
Eurycercus lamellatus	–	–	–	–	–	–	–	★	–	–	–	–	–	–	–	–	–	–	–	–	–	–	–	–	✪	◇	–	–	◇	–	–	–	–	–	–	★	–	–	14.29
Pleuroxus aduncus	–	◇	–	◇	◇	–	–	★	–	–	◇	–	–	◇	◇	–	–	–	–	–	–	–	–	–	–	–	◇	★	✪	◇	★	★	✪	✪	★	★	★	✪	54.29
Pleuroxus laevis	–	–	–	–	–	–	–	★	–	–	–	–	◇	–	–	–	–	–	–	–	–	–	–	◇	–	◇	–	–	–	–	–	–	–	–	–	–	–	–	11.43
Alonella excisa	–	–	–	◇	◇	–	–	–	–	–	◇	–	–	–	–	–	–	–	–	–	–	–	–	–	–	–	–	–	–	–	–	–	–	–	–	–	–	–	8.57
Alonella nana	–	–	–	◇	–	–	–	–	–	◇	–	–	–	–	–	–	–	–	–	–	–	–	–	–	–	–	–	–	–	–	–	–	–	–	–	–	–	–	5.71
Chydorus sphaericus	◇	◇	◇	★	★	–	–	★	–	◇	◇	◇	–	–	–	–	–	–	–	–	–	◇	–	◇	✪	◇	◇	★	✪	◇	✪	✪	✪	✪	✪	✪	★	✪	71.43
Alona costata	–	–	–	–	–	–	–	–	–	◇	◇	–	–	–	◇	–	–	–	–	–	–	–	–	–	–	–	–	–	–	–	–	–	–	–	–	–	–	–	8.57
Alona guttata	–	–	–	–	–	–	–	◇	–	–	–	–	–	–	–	–	–	–	–	–	–	–	–	–	–	–	–	–	–	–	–	–	–	–	–	–	–	–	2.86
Alona rectangula	–	–	–	–	–	–	–	◇	–	★	–	◇	◇	◇	◇	–	–	–	◇	◇	◇	◇	◇	◇	–	◇	–	★	★	–	◇	◇	–	◇	◇	–	◇	◇	60
Acroperus harpae	–	–	–	◇	–	–	–	–	–	–	◇	–	–	–	◇	–	–	–	◇		◇	–	–	◇	◇	–	★	◇	★	–	◇	◇	–	–	–	–	–	–	34.29
Graptoleberis testudinaria	–	–	–	–	–	–	–	–	–	◇	–	–	–	–	–	–	–	–	–	–	–	–	–	–	–	–	★	–	–	–	◇	–	–	–	–	–	–	–	8.57
Biapertura affinis	–	–	–	–	–	–	–	–	–	–	–	–	–	–	–	–	–	–	–	–	–	–	–	–	–	–	–	–	–	–	–	–	–	–	–	–	◇	–	2.86
Tretocephala ambiqua	–	–	–	–	–	–	–	–	◇	◇	–	–	–	–	–	–	–	◇	–	–	–	–	–	–	◇	–	–	–	–	–	–	–	–	–	–	–	–	–	11.43
Oxyurella tenuicaudis	–	–	–	–	–	–	–	–	–	–	–	–	–	–	–	–	–	–	◇	–	–	–	–	–	–	◇	◇	–	◇	–	◇	◇	–	◇	–	–	◇	◇	25.71

Study period	2002																														2003									
Months	March			April			May			June			July			August			September			October			November			December		January			February			March			
Stations/Species	1	2	3	1	2	3	1	2	3	1	2	3	1	2	3	1	2	3	1	2	3	1	2	3	1	2	3	1	2	1	2	3	1	2	3	1	2	3	f %
COPEPODA																																							
Acanthodiaptomus denticornis	–	–	–	–	–	–	✧	★	★	–	★	–	–	–	–	–	–	–	–	–	–	–	–	–	–	✧	–	★	✧	✧	★	★	✧	★	✧	★	★	✧	45.71
Macrocyclops albidus	–	–	–	–	–	–	–	–	–	–	–	–	–	–	–	–	–	–	–	–	–	–	–	–	–	✧	✧	–	★	–	–	–	–	–	–	–	–	–	8.57
Eucyclops macruroides	–	–	–	–	–	–	–	–	–	–	–	–	–	–	–	–	–	–	–	–	–	–	–	–	–	–	–	–	–	✧	–	–	★	★	✧	–	★	✧	17.14
Eucyclops macrurus	–	–	–	–	–	–	–	–	–	–	–	–	–	–	–	–	–	–	–	–	–	–	–	–	–	✧	–	✧	✧	–	★	–	–	★	✧	–	–	–	17.14
Eucyclops serrulatus	–	–	–	–	–	–	–	–	–	–	–	–	–	–	–	–	–	–	–	–	–	–	–	–	✧	–	–	–	–	–	–	–	–	–	–	–	–	–	2.86
Eucyclops speratus	–	✧	–	–	–	–	–	✧	–	–	–	–	–	–	–	–	–	–	–	–	–	–	–	–	✧	–	–	✧	–	–	–	–	–	–	–	–	–	–	11.43
Cyclops strenuus paternonis	–	–	–	–	✧	–	–	–	–	–	–	–	–	–	–	–	–	–	–	–	–	–	–	–	–	–	–	–	✧	–	★	–	–	★	✧	✧	★	✧	22.85
Megacyclops viridis	✧	–	–	–	✧	–	–	–	–	–	–	–	–	–	–	–	–	–	–	–	–	–	–	–	–	–	–	–	–	–	–	–	✧	–	–	–	–	–	8.57
Megacyclops gigas	–	–	–	–	–	–	–	–	✧	–	–	–	–	–	–	–	–	–	–	–	–	–	–	–	–	–	–	–	–	–	–	✧	–	–	–	–	–	–	8.71
Nitocra hibernica	–	–	–	–	–	–	–	–	–	✧	–	–	–	–	–	–	–	–	–	–	–	–	–	✧	–	–	–	–	–	–	–	–	–	–	–	–	–	–	5.71

Table 2. Distribution of zooplankton species found in the Lake Karamık according to months and stations (✿: abundant, ★: some, ✧:scarce) .

http://dx.doi.org/10.5772/63749

Species/authors	[1]	[46]	[80]	[41]	Present study
Rotaria neptunia		⬟			
Rotaria rotatoria		⬟			
Rotaria socialis		⬟			
Philodina megalotrocha		⬟			
Platyias quadricornis		⬟		⬟	⬟
Epiphanes senta	⬟				
Brachionus angularis		⬟			⬟
Brachionus calyciflorus	⬟	⬟			
Brachionus diversicornis		⬟			
Brachionus plicatilis		⬟			
Brachionus quadridentatus		⬟		⬟	⬟
Brachionus rubens		⬟			⬟
Brachionus urceolaris		⬟			
Keratella cochlearis	⬟	⬟			
Keratella tecta*					⬟
Keratella quadrata		⬟			⬟
Keratella tropica		⬟		⬟	
Notholca acuminata	⬟	⬟			⬟
Kellikottia longispina		⬟			
Anuraeopsis fissa		⬟			
Euchlanis dilatata		⬟			
Euchlanis incisa		⬟			⬟
Euchlanis meneta		⬟			⬟
Mytilina mucronata		⬟			⬟
Mytilina ventralis	⬟	⬟			⬟
Lophocharis rubens		⬟			
Lophocharis salpina		⬟			
Trichotria pocillum	⬟	⬟			⬟
Trichotria tetractis		⬟			⬟
Colurella adriatica		⬟			
Colurella colurus		⬟			

Species/authors	[1]	[46]	[80]	[41]	Present study
Colurella obtusa		●			
Colurella uncinata		●			
Colurella sp.	●				
Squatinella mutica		●			
Squatinella rostrum		●			
Squatinella sp.	●				
Lepadella ovalis		●			
Lepadella patella		●			
Macrochaetus collinsii		●			●
Lepadella ehrenbergi*					●
Lepadella ovalis*					●
Lepadella sp.	●				
Lecane elsa		●			●
Lecane inermis		●			
Lecane luna	●	●			●
Lecane nana		●			
Lecane ohioensis		●			●
Lecane stichaea		●			
Lecane furcata		●			
Lecane hamata		●			
Lecane lamellata		●			
Lecane bulla	●	●			●
Lecane clostrocerca		●			●
*Lecane cornuta**					●
Lecane lunaris		●			●
Lecane curvicornis*					●
Lecane quadridentata		●			●
Lindia truncata		●			
Scaridium longicaudum	●	●			
Monommata longiseta	●	●			
Eosphora najas		●			

Species/authors	[1]	[46]	[80]	[41]	Present study
Notommata copeus		●			
Pleurotrocha petromyzon		●			
Cephalodella auriculata		●			
Cephalodella catellina		●			
Cephalodella gibba	●	●			
Cephalodella sterea		●			
Cephalodella ventripes		●			
Trichocerca cylindrica	●				
Trichocerca iernis		●			●
Trichocerca similis*					●
Trichocerca tenuior		●			
Trichocerca weberi		●			
Ascomorpha ovalis		●			
Ascomorpha saltants		●			●
Synchaeta oblonga		●			
Synchaeta pectinata	●	●			●
Polyarthra dolichoptera		●			●
Polyarthra vulgaris	●	●			
Asplanchna girodi		●			
Asplanchna priodonta	●	●			
Asplanchna sieboldi		●			
Asplanchna sp.				●	
Dicranophorus grandis*					●
Encentrum saundersiae		●			
Testudinella patina		●			●
Pompholyx complanata		●			
Floscularia ringens		●			
Ptygura beauchampi		●			
Conochilus dossuarius		●			
Conochilus natans		●			
Conochilus unicornis		●			●

Species/authors	[1]	[46]	[80]	[41]	**Present study**
Hexarthra fennica		⬟		⬟	
Hexarthra mira		⬟			⬟
Hexarthra sp.	⬟				
Filinia longiseta		⬟			⬟
Filinia pejleri		⬟			
Filinia terminalis	⬟	⬟			⬟
Colletheca ornata		⬟			
Colletheca sp.*					⬟
Diaphanosoma lacustris			⬟		
Daphnia curvirostris				⬟	⬟
Daphnia longispina	⬟		⬟	⬟	⬟
Simocephalus exspinosus			⬟		⬟
Simocephalus vetulus	⬟		⬟		⬟
Ceriodaphnia dubia			⬟		
Ceriodaphnia quadrangula	⬟				⬟
Ceriodaphnia reticulata			⬟		⬟
Scapholeberis kingi	⬟		⬟		⬟
Moina branchiata			⬟		
Macrothrix laticornis	⬟				
Macrothrix rosea			⬟		
Bosmina longirostris*					⬟
Eurycercus lamellatus	⬟		⬟		⬟
Pleuroxus aduncus	⬟		⬟		⬟
Pleuroxus laevis	⬟				⬟
Alonella excisa	⬟				⬟
Alonella nana	⬟				⬟
Chydorus sphaericus	⬟		⬟		⬟
Alona costata	⬟		⬟		⬟
Alona guttata*					⬟
Alona rectangula			⬟		⬟
Acroperus harpae	⬟		⬟		⬟

Species/authors	[1]	[46]	[80]	[41]	Present study
Graptoleberis testudinaria	⬟		⬟		⬟
Biapertura affinis*					⬟
Tretocephala ambiqua			⬟		⬟
Oxyurella tenuicaudis	⬟		⬟		⬟
Acanthodiaptomus denticornis	⬟				⬟
Macrocyclops albidus	⬟				⬟
Eucyclops macruroides*					⬟
Eucyclops macrurus	⬟				⬟
Eucyclops serrulatus*					⬟
Eucyclops sperulus	⬟				⬟
Eucyclops sp.				⬟	
Cyclops strenuus paternonis	⬟				⬟
Cyclops sp.				⬟	
Megacyclops viridis	⬟				⬟
Megacyclops gigas*					⬟
Nitocra hibernica	⬟				⬟

Table 3. Checklist of zooplankton species studied during the present and earlier studies in Lake Karamık.

Among the species identified in the lake, *B. angularis, F. longiseta, Keratella quadrata* [19, 57], *B. longirostris* and *G. testudinaria* ([58–60] species grow in eutrophic conditions. According to QB/T *Rotifera* index, with the values of Karamık Lake with QB/T = 2.3 [46], Xochimilco Lake (Mexico) QB/T = 10 [81], Dojran Lake (Macedonia) QB/T = 1.6 [82], Beyşehir Lake (Beyşehir, Turkey) QB/T = 1.5–2 [30, 31], Gölcük Lake (Ödemiş, İzmir, Turkey) QB/T = 5 [48], Ladik Lake (Samsun, Turkey) QB/T = 5 [10], Abo Zaabal Lake (Cairo, Egypt) QB/T = 4 [83], Liman Lake (Kızılırmak delta, Turkey) QB/T = 2.5 [84] were shown as eutrophic. *Rotifera* species were used as indicator while identifying the trophic condition of the lake.

In this study, Karamık Lake was determined QB/T = 1.5. Finally, the lake showed mesotrophic property according to *Rotifera* index and eutrophic property in terms of *Rotifera* and *Cladocera* dominance. It is reported that the dominance of *Cladocera* and *Cyclopoid Copepoda* group of organisms in eutrophic waters [3], Cyclops strenuus *Cyclopoid* copepod dominance in freshwater widely in oligotrophic waters and rarely in shallow eutrophic waters in Japan [85]. *B. angularis, B. quadridentatus, K. qaudrata, P. dolichoptera, F. longiseta, C. qaudrangula, B. longirostris, C. sphaericus, Daphnia longispina* and *Cyclops strenuus paternonis* species identified in this study are the indicators of mesotrophic-eutrophic.

While *Keratella cochlearis, Keratella quadrata* and *Polyarthra vulgaris* species were reported as dominant species in the study made in the lake in 1984, the dominance of these little herbivorous zooplanktons (*Rotifera* and *Cladocera*) in the Lake Karamık was presented as the proof that this lake was eutrophic [1]. After the closure of the paper mill around Karamık Lake, becoming hypereutrophic of the lake was prevented. Existing quantitatively more of the zooplankton species than zooplankton groups depends on high level of food, breeding success of the *Rotifera* species and above all restrain of the increase in *Cladocera* and *Copepoda* by fish [46]. Consequently, the dominance of the number *Rotifera* and *Cladocera* species and also being in the large form of the *Cladocera* group of organisms show that the lake has turned into eutrophic condition. In this research, the shallow eutrophic lakes in Turkey were reported, and the zooplankton composition in shallow Lake Karamık was studied in detail, and zooplanktonic organisms of the eutrophic lakes were presented. Finally, this study will be useful contributions to the notice of zooplankton fauna of eutrophic lakes and of Turkey's biodiversity.

Acknowledgements

The author is grateful to Eğirdir Fisheries Research Institute for its support in taking samples from the lake, to Aquaculture Engineer Rahmi UYSAL from the technical staff of the Institute, to Aquaculture Engineer M.Sc. Abdulkadir YAĞCI for his support in bringing samples from the lake and designing the shape of the study area.

Author details

Meral Apaydin Yağci

Address all correspondence to: meralyagci@gmail.com

Fisheries Research Institute, Eğirdir, Isparta, Turkey

References

[1] Gündüz E: Determination of zooplankton species and the effects of pollution on zooplankton in Karamuk and Hoyran Lakes [thesis]. Hacettepe University Graduate School of Science and Engineering. Ankara; 1984.

[2] Baloch WA, Jafri SIH, Soomro AN: Spring zooplankton composition of Rawal Lake, Islamabad. Sindh Univ Res J (Sci Ser). 2005; 37: 41–46.

[3] Altındağ A, Buyurgan Ö, Kaya M, Özdemir E, Dirican S: A survey on some physico-chemical parameters and zooplankton structure in Karaman Stream, Antalya, Turkey. J Anim Vet Adv. 2009; 8: 1710–1716.

[4] Yıldız Ş, Özgökçe, MS, Özgökçe F, Karaca İ, Polat, E: Zooplankton composition of Van Lake coastline in Turkey. Afr J Biotechnol. 2010; 9: 8248–8252.

[5] Apaydın Yağcı M: Seasonal zooplankton community variation in Karataş Lake, Turkey. Iran J Fish Sci. 2013; 12: 265–276.

[6] Pinto-Coelho R, Pinel-Alloul R, Méthot G, Havens KE: Crustancean zooplankton in lakes and reservoirs of temperate and tropical regions: Variation with trophic status. Can J Fish Aquat Sci. 2005; 62: 348–361.

[7] Sukumaran PK, Das AK: Distribution and abundance of rotifers in relation to water quality of some tropical reservoirs. Indian J Fish. 2004; 51: 295–301.

[8] Kamaladasa AI, Jayatunga YNA: Composition, density and distribution of zooplankton in south west and east lakes of Beira Lake soon after the restoration of south west lake. Cey J Sci. 2007; 36: 1–7.

[9] Ezekiel EN, Ogamba EN, Abowei JFN: The zooplankton species composition and abundance in Sombreiro River, Niger Delta, Nigeria. Asian J Agric Sci. 2011; 3: 200–204.

[10] Apaydın Yağcı M, Yılmaz S, Yazıcıoğlu O, Polat N: The zooplankton composition of Lake Ladik (Samsun, Turkey). Turk J Zool. 2015; 39: 652–659.

[11] Güher H: A faunistic study on the freshwater Cladocera (Crustacea) species in Turkish Thrace (Edirne, Tekirdağ, Kırklareli). Turk J Zool. 2000; 24: 237–243.

[12] Rajashekhar M, Vijaykumar K, Parveen Z: Zooplankton diversity of three freshwater lakes with relation to trophic status, Gulbarga District, North-East Karnataka, South India. Int J Syst Biol. 2009; 1: 32–37.

[13] Gökçe D, Özhan Turhan D: Evaluation of vertical and horizontal changes in community structure of zooplankton in a deep dam lake. Turk J Zool. 2014; 38: 11–22.

[14] Segers H: Global diversity of Rotifers (Rotifera) in freshwater. Hydrobiologia. 2008; 595: 49–59.

[15] Hra M: Seasonal and spatial distribution of Brachionus (Pallas, 1966; Eorotatoria: Monogononanta: Brachionidae), a bioindicator of eutrophication in lake El-Manzalah, Egypt. Biol Med. 2011; 3: 60–69.

[16] Špoljar M, Tomljanović T, Lalić I: Eutrophication impact on zooplankton community: a shallow lake approach. Holist Approach Environ. 2011; 1: 131–142.

[17] Baloch WA, Suzuki H: Summer zooplankton composition, vertical distribution and biomass in Lake Ikeda, Southern Kyushu, Japan. Sindh Univ Res J (Sci Ser). 2009; 41: 35–40.

[18] Shah JA, Pandit AK: Seasonal succession of crustacean zooplankton in Wular Lake of the Kashmir Himalaya. Arch Biol Sci Belgrade. 2013; 63: 1063–1068.

[19] Geng H, Xie P, Deng D, Zhou Q: The rotifer assemblage in a shallow, eutrophic Chinese lake and its relationships with cyanobacterial blooms and crustacean zooplankton. J Fresh Ecol. 2005; 20: 93–100.

[20] Sarma SSS, Nandini S, Gulati RD: Life history strategies of cladocerans: comparison of tropical and temperate taxa. Hydrobiologia. 2005; 542: 315–333.

[21] Gutkowska A, Paturej E, Kowalska E: Rotifer trophic state indices as ecosystem indicators in brackish coastal waters. Oceanologia. 2013; 55: 887–889.

[22] Xiong J, Mei X, Liu J: Comparative studies on community structure, biodiversity of plankton and zoobenthos in four lakes of different trophic states in China. Asian Fish Sci. 2003; 16: 361–372.

[23] Maqbool A, Sulehria AQK, Ejaz M, Hussain A: Density, diversity and abundance of Copepods in a pond. Biologia. 2014; 60: 57–62.

[24] Chen G, Dalton C, Taylor D: Cladocera as indicators of trophic state in Irish lakes. J Paleolimnol. 2010; 44: 465–481.

[25] Illyová M. Ecological integrity of river-floodplain system-assessment by planktonic crustaceans surveys (Branchiata: Branchiopoda). XVI. Konferencia (CLA A SLS); 25–29 June 2012; Jasná: pp. 50–53.

[26] Segers H: Rotifera: Monogononta. In: Yule CM, Yong HS, editors. Handbook of freshwater invertebrates of the Malaysian region. Kuala Lumpur, Academy of Sciences of Malaysia; 2004. pp. 106–120.

[27] Ustaoğlu MR: An updated zooplankton biodiversity of Turkish inland waters. J Limnol Freshw Fish Res. 2015; 1 (3): 151–159.

[28] Baiao C, Boavida MJ: Rotifers of Portuguese reservoirs in River Tejo catchment: relations with trophic state. Limnetica. 2005; 24: 103–114.

[29] Dumont HJ. Zooplankton and the science of biogeography: the example of Africa. In: Kerfoot, W, editors. Handbook of evolution and ecology of zooplankton communities. Special Symposium Volume 3 American Society of Limnology and Oceanography, University Press of New England Hanover, New Hampshire and London, England; 1980. pp. 685–696.

[30] Özdemir N, Atamanalp M. Rotifers in Aquatic Toxicology, Ist Symposium on Fish Introduction and Reservoir Management, Antalya: 2006. p. 509-516.

[31] Altındağ A, Yiğit S: Beyşehir Gölü Zooplankton Faunası ve Mevsimsel Değişimi. G.Ü. Gazi Eğitim Fakültesi Dergisi. 2004; 24: 217–225.

[32] Didinen H, Boyacı YÖ: Lake Beyşehir Monthly Distribution of Zooplankton Species. Journal of Natural and Applied Science. 2014; 18: 91-98.

[33] Apaydın Yağcı M, Yağcı A, Bilgin F: Study on composition and abundance of zooplankton assemblages in Eğirdir Lake (Isparta, Turkey). Iran J Fish Sci. 2014; 13: 834–855.

[34] Yıldız Ş, Altındağ A, Ergönül MB: Seasonal fluctuations in the zooplankton composition of a Eutrophic Lake: Lake Marmara (Manisa, Turkey). Turk J Zool. 2007; 31: 121–126.

[35] Ustaoğlu MR, Akyürek M. Zooplankton of Lake Akşehir. XII. National Biology Congress. 6-8 July 1994; Edirne.

[36] Altındağ A: Seasonal Variations of the Zooplanktonic Community of Lake Akşehir. Ege University, Faculty of Fisheries, Journal of Fisheries and Aquatic Sciences. 1997; 14: 57-69.

[37] Ustaoğlu MR: Zooplankton (metazoa) of the Karagöl (Yamanlar, İzmir-Turkey). Biol Gallo-hell. 1986; 12: 273–281.

[38] Emir N: Zooplankton community structure of Çavuşcu and Eber Lakes in Central Anatolia. Acta Hydrochim Hydrobiol. 1994; 22: 280–288.

[39] Güher H, Erdoğan S, Kırgız T, Çamur-Elipek B: The dynamics of zooplankton in National Park of Lake Gala (Edirne-Turkey). Acta Zool Bulg. 2011; 63: 157–168.

[40] Altındağ A, Yiğit S, Ergönül MB: The zooplankton community of Lake Mogan, Turkey. J Freshw Ecol. 2007; 22: 709–711.

[41] Kazancı N, Girgin S, Dügel M, Oğuzkurt D, Mutlu B, Dere Ş, Barlas M, Özçelik M. Köyceğiz, Beyşehir, Eğirdir, Akşehir, Eber, Çorak, Kovada, Yarışlı, Bafa, Salda, Karataş, Çavuşçu, Small and Large Menderes Delta, Güllük Reed Bed, Karamuk Marsh Limnology, Environmental Quality and Biodiversity. Turkey Inland Waters Research Series. IV, Ankara; 1999. 371p.

[42] Apaydın Yağcı M: Seasonal variations in zooplankton species of Lake Gölhisar, a shallow lake in Burdur, Turkey. Pak J Zool. 2014; 46: 927–932.

[43] Bozkurt A: Zooplankton of Yenişehir Lake (Reyhanlı, Hatay). E.U. Journal of Fisheries & Aquatic Sciences. 2006; 23: 39-43.

[44] Güher H, Kırgız T, Çamur B, Güner U: A study on zooplankton organisms community structures of Lake Terkos (İstanbul-Turkey). Pak J Biol Sci. 2004; 7: 566–570.

[45] Bekleyen A, Taş B: Zooplankton fauna of Çernek Lake (Samsun). Ekoloji. 2008; 17: 24–30.

[46] Emir N, Demirsoy A: Seasonal variation of the zooplanktonik organizms of Karamuk Lake. Turk J Zool. 1996; 20: 137–144.

[47] Serteser A, Acar H: Aquatic macrophytes and soil features on Karamik Lake coastal ecosystem in Afyonkarahisar (Turkey). J Biol Environ Sci. 2014; 8: 143–150.

[48] Saygı Y, Gündüz E, Demirkalp FY, Çağlar SS: Seasonal patterns of the zooplankton community in the shallow, brackish Liman Lake in Kızılırmak Delta, Turkey. Turk J Zool. 2011; 35: 783–792.

[49] Özdemir Mis D, Ustaoğlu MR: Investigations on zooplankton of Gölcük Lake (Ödemiş, İzmir). Eu J Fish Aquat Sci. 2009; 26: 19–27.

[50] Yasan AB. Determination of Trophic Status of Eber (Afyon) Lake [thesis]. Ankara University Graduate School of Natural and Applied Sciences Department of Biology. Ankara; 2007.

[51] Yiğit S: Seasonal fluctuation in the Rotifer fauna of Kesikköprü Dak Lake (Ankara, Turkey). Turk J Zool. 2002; 26: 341–348.

[52] Güher H: A checklist of zooplankton (Rotifera, Copepoda, Cladocera) of European Turkey inland waters. Ege J Fish Aqaut Sci. 2014; 31: 221–225.

[53] Ruttner-Kolisko A. Handbook of plankton rotifers, biology and taxonomy. Die Binnengewasser. XXVI/I. Stuttgart; 1974. 144p.

[54] Inaotombi S, Gupta PK, Mahanta PC: Influence of abiotic factors on the spatio-temporal distribution of rotifers in a subtropical lake of Western Himalaya. Water Air Soil Pollut., 2016; 227:50.

[55] Bozkurt A, Tepe Y: Zooplankton composition and water quality of Lake Gölbaşı (Hatay-Turkey). Fresenius Environ Bull. 2011; 20: 166–174.

[56] Bozkurt A, Akın Ş. Zooplankton fauna of Yeşilırmak (between Tokat and Blacksea), Hasan Uğurlu and Suat Uğurlu Dam Lakes. Turk J Fish Aquat Sci. 2012; 12: 777–786.

[57] Imoobe TOT, Adeyinka ML: Zooplankton-based assessment of the trophic state of a tropical forest river in Nigeria. Arch Biol Sci Belgrade. 2009; 61: 733–740.

[58] Saksena, DN: Rotifers as indicators of water quality. Acta Hydrobiol. 1987; 15: 481–485.

[59] Makarewicz JC: A lakewide comparison of zooplankton biomass and its species composition in Lake Erie. J Great Lakes Res JGLRDE, 1993; 19: 275–290.

[60] Bos DG, Cumming BF: Sedimentary cladoceran remains and their relationship to nutrients and other limnological variables in 53 lakes from British Columbia, Canada. Can J Fish Aquat Sci. 2003; 60: 1177–1189. DOI: 10.1139/f03-097

[61] Sendacz S, Caleffi S, Santos-Soares J: Zooplankton biomass of reservoir in diff erent trophic conditions in the state of Sao Paulo. Braz J Biol. 2006; 66: 337–350.

[62] Sládeček V: Rotifers as indicators of water quality. Hydrobiology. 1983; 100: 169–201.

[63] Gannon JE, Stemberger RS: Zooplankton (especially crustaceans and rotifers) as indicators of water quality. Trans Am Microsc Soc. 1978; 97: 16–35.

[64] Dorak Z: Zooplankton abundance in the lower Sakarya River Basin (Turkey): impact of environmental variables. J Black Sea/Mediterranean Environ. 2013; 19: 1–22.

[65] Ejsmont-Karabin J: The usefulness of zooplankton as lake ecosystem indıcators: rotifer trophic state index. Polish J Ecol. 2012; 60: 339–350.

[66] Aksun FY: The Reproduction Biology of Pike (*Esox lucius* L., 1758) in Karamık Lake. DOĞA Turk. Zool. Journal. 1987a; 11: 67-75.

[67] Aksun FY: The Growth Features and Growth Rates of Pike (*Esox lucius* L., 1758) in Karamık Lake. DOĞA Turk. Zool. Journal. 1987b; 11: 76-86.

[68] Çubuk H, Balık İ, Özkök R, Uysal R, Yağcı A. Determination of Biological Properties of Fisheries and Economic Fish Species in Lake Karamık. Project Number: Tagem-Haysüd-2-00-17-03.Fisheries Research Institute, Eğirdir, Isparta, Turkey; 2004.

[69] Dussart B. Les Copédes des eaux continentales d'Europe occidentale. Tome I. Calanoides et Harpacticoides. Paris, France: N. Boubee et cie.1967. 500p.

[70] Dussart B. Les Copédes des eaux continentales d'Europe occidentale. Tome II. Cyclopoides et Biologie. Paris, France: N Boubee et cie. 1969. 292p.

[71] Koste W. Rotatoria, Die Radertiere Mitteleuropas ein Bestimmungswerk, begrundet von Max Voigt Uberordnung Monogononta,I Textband and II Textband, Gebruder Borntraeger, Berlin, Stuttgart. 1978. 234p. and 673p.

[72] Negrea S. Fauna republici socialiste Romania. Crustacea, Cladocera. Bucharest, Romania: Acedemia Rep. Soc. Romania. 1983. 367p.

[73] Korovchinsky NM: Sididae and Holopedidae (Crustacea: Daphniiformes), Guides to identification of the microinvertebrates of the continental waters of the World. The Netherlands, SPB Academic Publishers. 1992. 82p.

[74] Segers H: The Lecanidae (Monogononta). Vol. 2, Guides to the identification of the microinvertebrates of the continental waters of the World. Coordinating editor: HJF Dumont. Gent Univ., Belgium. The Netherlands, SPB Academic Publishers; 1995. 226p.

[75] Smirnov NN: Guides to the identification of the microinvertebrates of the continental waters of the world. Cladocera: the Chydorinae and Sayciinae (Chydoridae) of the World No. 11. The Hague, the Netherlands, SPB Academic Publishers; 1996. 197p.

[76] Nogrady T, Segers H. Asplanchnidae, Gastropodidae, Lindiidae, Microcodidae, Synchaetidae, Trochosphaeridae and Filinia. In: Dumont HJ, editor. Guides to the identification of the microinvertebrates of the continental waters of the world, Vol. 6. Leiden, the Netherlands, Backhuys Publishers BV; 2002. 264p.

[77] Soyer J: Bionomie benthique du plateau continental de la côte catalane française. III. Les peuplements de Copépodes Harpacticoides. Vie et Milieu Ser (B) Océanogr., 1970; 21: 337-511.

[78] Ustaoğlu MR: A checklist for zooplankton of Turkish inland waters. Ege Univ J Fish Aquat Sci. 2004; 21: 191–199.

[79] Ustaoğlu MR, Altındağ A, Kaya M, Akbulut N, Bozkurt A, Özdemir Mis, Atasagun S, Erdoğan S, Bekleyen A, Saler S: A checklist of Turkish rotifers. Turk J Zool. 2012; 36: 607–622.

[80] Gündüz E: A Checklist of Cladoceran Species (Crustacea) Living in Turkish Inland Waters. Turk. J. Zool., 1997; 21: 37-45.

[81] Nandini S, Ramírez García P, Sarma SSS: Water quality indicators in Lake Xochimilco, Mexico: zooplankton and Vibrio cholerae. J Limnol. 2016; 75: 91–100.

[82] Tasevska O, Jersabek CD, Kostoski G, Guševska D: Differences in rotifer comminities in two freshwater bodies of different degree (Lake Ohrid and Lake Dojran, Macedonia). Biologia. 2012; 67: 565–572.

[83] El-Bassat RA, Taylor WD: The zooplankton community of Lake Abo Zaabal, a newly-formed mining lake in Cairo, Egypt. Afr J Aquat Sci. 2007; 32: 185–192.

[84] Demirkalp FY, Saygı Y, Çağlar SS, Gündüz E, Kılınç S: Limnological assesment on the brackish shallow Liman Lake from Kizilirmak Delta (Turkey). J Anim Vet Adv. 2010; 9: 2132–2139.

[85] Makino W, Ban S: Diel changes in vertical overlap between *Cyclops strenuus*—Copepoda; Cyclopoida/and its prey in oligotrophic Lake Toya, Hokkaido, Japan. J Mar Syst. 1998; 15: 139–148.

Chapter 6

Phycoremediation of Eutrophic Lakes Using Diatom Algae

Marella Thomas Kiran,
Mallimadugula Venkata Bhaskar and
Archana Tiwari

Additional information is available at the end of the chapter

http://dx.doi.org/10.5772/64111

Abstract

Eutrophication as a result of human intervention has led to severe deterioration of fresh water habitats. Due to population growth, industrialization and uncontrolled use of fertilizers led to excess nutrient runoff entering into rivers and lakes; this has caused reduction in water quality and abnormal changes in ecosystem structure and function. A solution to this cultural eutrophication is an urgent necessity since nutrient accumulation renders controlling eutrophication more difficult over time. Using algae for reduction of nutrients is a unique technology, which utilizes the enormous potential of microalgae in restoring water quality. This has a huge potential in urban lakes where there is an urgent need to use such technologies in combination with existing ones to speed up the process to reduce the formation of hypereutrophic lakes and dead zones in oceans. In this book chapter, we explore the enormous potential of diatoms as cost-effective, efficient and eco-friendly remedy for complex problems related to eutrophication. We report the case studies on using diatom-based technology. This will give us a new insight into microalgae-based lake remediation strategies, which can significantly reduce the cost, manpower needed and negative environmental impacts involved in existing technologies.

Keywords: diatoms, eutrophication, phycoremediation, Nualgi

1. Introduction

Many human activities are polluting freshwater ecosystems, modifying the structure of aquatic communities and thereby disrupting the functional continuum of river and lake systems. Indeed, anthropogenic pollution of freshwater ecosystems by the addition of organic matter and nutrients is an increasing phenomenon that affects many lakes and rivers worldwide [1]. Physicochemical alterations caused by contaminated water and biological wastes often involve increases in inorganic nutrients (ammonium, nitrate, nitrite, phosphate) and suspended organic solids, a decrease in dissolved oxygen, and a settlement of suspended organic matter settling on the lake bottom. Therefore, there is an urgent need to explore new eco-friendly, cost-effective strategies to mitigate nutrient input into waterways and also to remove nutrients from waterways.

Nutrient contamination of surface waters has led to widespread excessive algae growth, a process known as eutrophication. Eutrophication can lead to fish kills through oxygen depletion or the growth of toxic dinoflagellates that produce neurotoxins harmful to fish and humans [2–4]. Eutrophication also can cause taste and odour issues that create expensive problems for municipalities that rely on surface water for their drinking water and individual households depending on groundwater [1].

Benthic diatoms are the dominant algal community in water bodies and they contribute significantly to nutrient removal and dissolved oxygen levels in water. They also form the basis of benthic food web in water bodies. Diatom algae contribute up to 40% of primary production in lakes and oceans, which is more than that of all the tropical rain forests on earth. Diatoms play an important role as a major carbon carrier to Deep Ocean to be one of the major contributors to the "biological carbon pump".

Diatoms are microscopic plants, which use nitrates and phosphates to grow along with other nutrients such as silica, iron, copper, molybdenum, etc.; they use CO_2 and produce O_2 and they can also accumulate heavy metals, so by triggering the growth of these algae, many problems related to lake pollution can be solved. Growth of diatoms also reduces the growth of harmful algae such as blue green algae (BGA) [5]. Diatoms are important primary producers in streams, lakes and wetlands [6]. The main source of energy in streams was once thought to be detritus from terrestrial origin but later research showed that primary production by algae was important in many streams. Diatoms are now predicted to be the primary energy source in many streams [7]. Diatoms are also known to be important sources of energy for invertebrates in some headwater streams and even dominant, primary producers in many shallow lakes and ponds. In wetlands, diatoms are significant primary producers because of their high turnover rate. In addition to primary producers, diatoms are chemical modulators in aquatic ecosystems [8]. They transform many inorganic chemicals into organic forms. Diatoms are primary harvesters of inorganic phosphorus and nitrogen in stream spiralling in lake littoral modulation of influxes and in wetlands. Diatoms on surface sediments and plants are considered to be important sinks for nutrients before release into the water.

Diatoms as indicators of lake water quality were well studied by many researchers but diatoms also play a significant role in maintaining the water quality [9–11], so using diatom algae for nutrient removal is novel and cost-effective method of water treatment. The main bottleneck in using only diatom for nutrient removal is to trigger only diatom growth instead of other algae, so to solve this problem, the main solution is to use the silica as the nutrient, which is absolutely required by diatoms for their growth.

Phycoremediation is defined as the use of algae to remove pollutants from the environment or to render them harmless [12]. Phycoremediation has evolved from the early work done by Oswald and Gotaas [13] for the use of microalgae for tertiary treatment of municipal wastewater to many other applications in which microalgae are cultivated and utilized for specific bioremediation needs. The use of microalgae for the treatment of municipal wastewater has been the subject of research and development for several decades.

Considering excess nitrate and phosphate as a resource and not as a pollutant is a key in unlocking the problems related with nutrient pollution. Solar energy can be harnessed to grow algal biomass on wastewater nutrients and this biomass can be a source on which fish can be grown; this could provide a holistic solution to nutrient management problems. Nutrient runoff from agriculture and sewage constitutes a major component of wastewater generated everyday in both developed and developing nations such as United States and India. Excess N and P fertilizers and untreated or partially treated sewage find its way to the water bodies resulting in enrichment of nutrients, leading to eutrophication and formation of dead zones. One of the highly significant and rapidly developing methods for nutrient removal is the use of photosynthesis by growing algae. The use of algae for municipal wastewater treatment in ponds is well established [14, 15]. Algae growth in wastewater treatment ponds contributes to treatment mainly through dissolved oxygen production and nutrient assimilation. The release of oxygen from water during photosynthesis provides aerobic microbiological waste oxidation and the absorption of carbon dioxide which also accompanies photosynthesis.

The use of microalgae for the treatment of municipal wastewater has been the subject of research and development for several decades. In the early1950s, the first research on using microalgae for wastewater treatment was started. It was demonstrated that algae-based wastewater treatment could remove the nutrients (e.g. N and P) from settled domestic sewage more efficiently than traditional activated sewage process, indicating a great potential of algae-based wastewater treatment system. The result of such effort is that some commercial technologies and processes are available in the market such as the Advanced Integrated Wastewater Pond Systems (AIWPS) Technology commercialized by Oswald and Green [16], LLC, in the United States. Sewage contains mainly N and P, so this technology of Oswald's will aptly suite for excess nutrient removal in any water body but the main obstacle is to grow a particular type of algae in a controlled way. This can be achieved by using our technology.

1.1. Phycoremediation

Phycoremediation in a much broader sense is the use of macroalgae for the removal or biotransformation of pollutants, including nutrients and xenobiotics from wastewater and CO_2 from air. Algae can fix carbon dioxide by photosynthesis and remove excess nutrients efficiently at minimal cost; in addition, photosynthetically produced oxygen can relieve biological oxygen demand (BOD) in waste water.

Phycoremediation comprises several applications such as oxygenation of the atmosphere, nutrient removal from municipal wastewaters and effluents rich in organic matter, nutrient and xenobiotic compounds removal by biosorption using algae, treatment of acidic and metal wastewaters, CO_2 sequestration, transformation and degradation of xenobiotics and biosensing of toxic compounds by algae.

1.2. Diatoms and aquatic ecosystems

Phytoplankton community composition is highly dependent on the quantity and ratio of macro- and micronutrients in aquatic ecosystems. There are many examples of taxonomic shifts due to the relative supply of silica (Si) versus other nutrients (e.g. nitrogen (N) and phosphorus (P)). Bacillariophytes, or diatoms, are fast-growing phytoplankton that utilize dissolved silicate (SiO_4) to make their siliceous-armoured skeletal frustules [17]. In marine systems, diatoms require a particulate cell N/Si ratio of ~1 for balanced growth. Other phytoplankton species, such as dinoflagellates, cyanophytes, haptophytes and raphidophytes, do not utilize Si. If silicate is limiting, these other phytoplankton are capable of outcompeting diatoms despite generally slower growth rates [18]. Therefore, by 'fertilizing' waters those are depleted in Si relative to other macronutrients, such as with high Si-content solutions, the potential exists to shift the phytoplankton community to diatom dominance.

Diatoms are a widespread, diverse group of microalgae found in all aquatic systems. They represent a major component at the base of the marine food web, responsible for up to 50% of total lake and oceanic primary production [7, 19] and 25% of all oxygen produced on the planet. It also absorbs 23.5% of carbon dioxide generated on the planet. Diatoms can be found from the poles to the tropics, vary in size (2–200 μm), shape (centric, pennate), and can exist as single cells, colonies or chains [20]. Diatoms are opportunistic, generally exhibiting high growth rates and blooming rapidly when nutrient and light conditions are favourable [21]. Similarly, blooms can end as quickly when the diatoms have utilized all available nutrients and are either grazed upon (supporting higher trophic levels) or sink rapidly (driving the carbon pump). Diatoms require less light than other algae [22]; since their silica shells are transparent, they grow even on cloudy or rainy days. Diatoms can even dominate under nutrient-limiting conditions; in one example, a diatom species was shown to outcompete non-N-fixing cyanobacteria under low nitrate concentrations in a eutrophic lake [23]. These factors make most species of diatoms effective nutrient "sponges". This combination of the diatom's abilities makes them an ideal organism for water remediation practices.

1.3. Diatom algae for nutrient removal

Growing microalgae/phytoplankton such as diatom algae in the sewage will enable the nutrients in the sewage to be consumed and the oxygen produced will satisfy the BOD and chemical oxygen demand (COD) and provide oxygen to fish. Phytoplankton is the natural food for fish and diatoms are the best group of phytoplankton. Thus, polluted lakes will become clean and have plenty of fish. About 50% of the photosynthesis on Earth takes place in water—lakes and oceans and diatom algae account for about 50% of the algae in water bodies. Treated water with BGA and green algae cannot be released into public water bodies. Diatoms assimilate a significant amount of nutrients because they require high amounts of nitrogen and phosphorus for the synthesis of proteins (45–60% of microalgal dry weight), nucleic acids and phospholipids. Nutrient removal can also be further increased by NH_3 stripping or P precipitation due to the rise in the pH associated with photosynthesis.

1.4. Advantages of treating waste water with diatoms

Diatom algae are the most prolific microalgae in nature and they grow in all ponds, lakes, rivers, oceans, aquariums, etc. They are responsible for about 20–25% of all photosynthesis on Earth [24]; this is more than the share of tropical rain forests (~18%) and agriculture (8%). Diatoms can consume N and P faster than other algae. Diatoms can consume all forms of N such as nitrate, nitrite, urea and ammonia. Diatoms are best sequesters of CO_2, so they can release more oxygen.

1.5. Is diatom-based phycoremediation important?

Diatoms have an absolute requirement for significant amounts of silicon. Total algal biomass cannot be limited by Si, but its availability will shape phytoplankton communities. Ryther and Officer [25, 26] suggest that the eutrophication of waters by domestic wastes relatively poor in Si could lead to Si depletion and the elimination of diatoms from the phytoplankton communities. This process has been documented in the Laurentian Great Lakes [27]. Centric diatoms have been classed as the most desirable phytoplankton in coastal and fresh waters because they are important in aquatic food chains, they do not form noxious surface blooms and they are not toxic. Marine diatoms often have high growth rates and some freshwater diatoms have been shown to outcompete other algal groups for both N and P when adequate Si is available [5, 28, 29]. Therefore, the availability of Si can make diatom consume N and P at a faster rate than other undesirable species such as blue green algae and flagellates. Dissolved silica becomes available in waters primarily through the weathering of silicate rocks. Domestic wastes have low concentrations of Si relative to N or P [25], and the relative proportion of Si to either N or P is very low compared to the requirements of diatoms and the relative abundance of these elements in natural water. The proportion in which nutrients are loaded to a system can exert a strong influence on which algal species will thrive [30]. In this regard, the apparent growing preoccupation with nitrogen and phosphate in the literature may be counterproductive. A balanced approach emphasizing the interplay of various nutrients including the trace metals in shaping phytoplankton communities and their response to enrichment is required. If we can manage the species composition of eutrophic systems to promote the growth of algal species such as diatoms that, in turn, increase the secondary productivity of valuable food species, then we will have solved the important nutrient limitation riddle.

1.6. Mechanism to trigger diatoms in open waters

Growing one type of algae in open waters is a complicated process, but diatoms have a distinctive advantage because of their absolute requirement for silica. Therefore, by using this advantage we can trigger diatoms in open waters especially in fresh water ecosystem where silica concentration is less. Taking this advantage into consideration, we have developed a nanosilica-based micronutrient mixture called "Nualgi", which has nanosilica as its major constituent along with iron and nine other trace metals. The silica becomes both the carrier for other nutrients and the nutrient by itself. It is in a water-dispersible particulate form. Nualgi because of its nanosize is able to pervade very small spaces in the subsurface and remain suspended in water, allowing the particles to travel farther than larger, macro-sized particles there by increasing the bioavailability of the nutrients for easy absorption by microalgae and achieve wider distribution. In water, Nualgi causes diatom algae to bloom, though any pond, lake, estuary or coastal water has many species of organisms, only diatoms require silica and they consume Nualgi rapidly and bloom. In laboratory experiments with pure marine diatom cultures, highest biomass concentration and biomass productivity were attained in both *C. clostridium* and *C. fusiformis* in cultures grown in Nualgi containing medium; these values were almost double than in the popular f/2 medium [31] (**Figure 1**).

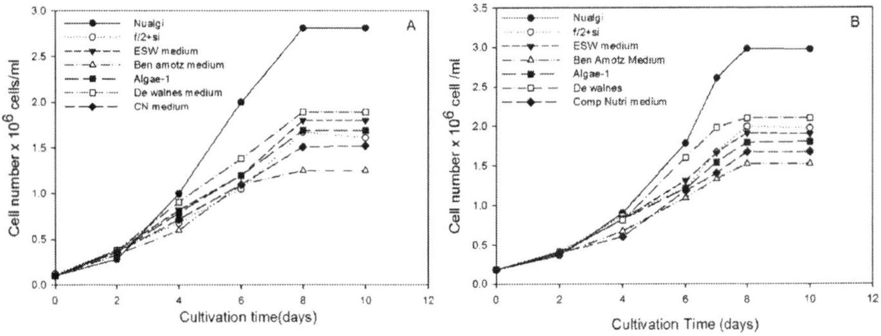

Figure 1. Effect of six different mediums on the growth of *C. clostredium* (A) and *C. fusiformis* (B) when compared with Nualgi-containing medium.

From Laboratory trails with eutrophic fresh water from Hussain Sagar Lake, Nualgi triggered diatom growth by not only triggering an increase in the number of diatom cells but also increasing diatom diversity. In samples without Nualgi, six different species of diatoms were identified in which *Cyclotella meneghiniana, Gomphonema lanceolatum* and *Nitzschia palea* were the abundant species, whereas in samples with Nualgi addition *Achnanthidium exiguum, Navicula cryptocephala, Cymbella tugidula, Navicula gracilis* and *Pleurosigma elongatum* were the dominant species with a total of 30 different diatom species. The dominant species in samples with no Nualgi addition were identified as pollution-tolerant species and in Nualgi-added samples dominant species are less pollution-tolerant species; this change in diatom species diversity clearly indicates the effect of Nualgi addition in nutrient reduction. In field conditions with Nualgi, N-removal percentage was 95.1% and P removal of 88.9% was achieved over a period of 10 days. COD and BOD reduction was also significant with 91 and 51%, respectively (Thomas, unpublished). Nualgi as an efficient tool to trigger diatom growth is well proven in both laboratory and field trials. Therefore, by using this product we want to propose a new strategy to mitigate excess nutrients by growing diatom algae in large fresh and brackish water bodies [32].

1.7. Bioremediation versus phycoremediation

There is a general understanding that bioremediation is akin to phycoremediation. In actual, they are very different and bioremediation is at best a subprocess of phycoremediation.

In bioremediation, bacteria are dosed into the water bodies. These break down the organic matter in the sewage, dead algae and weeds. Bacteria consume oxygen, so aerators may have to be used to provide the oxygen required. The organics are removed to a certain extent but the dead bacteria sediment and accumulate. Thus, most of the nutrients remain in the lake. Bacteria are cultured and dosed periodically. This is quite expensive. In nature, bacteria help in the digestion of food even in human digestive system and in sewage treatment plants (STPs). This is being copied. Mechanical aerators result in the release of CO_2 at the power plants or from the diesel engines. Similarly, STPs produce sludge, which is difficult to dispose off.

In phycoremediation, nutrient enrichment with Nualgi causes the native diatoms present in all water bodies to grow. The oxygen produced by diatoms causes the native bacteria to grow and these work in the same way as the bacteria dosed in bioremediation solutions. Diatoms release pure oxygen during photosynthesis, resulting in increased DO levels; this will lead to cascading improvement in water quality, aquatic life and biodiversity. Diatoms are consumed by zooplankton and fish and thus exit the water as fish biomass. Very few diatoms die and fall into the lake bed. Diatoms grow with a small dose nutrients and since fish consume them, the income from the sale of fish will recover most of the cost of Nualgi used. The cost and quantum of dosage are much less than the dosing of bacterial strains under bioremediation. Phycoremediation restores the natural food chain in the lake and this is the best way to remove nutrients. The native bacteria break down the nutrients and diatoms help remove them from the water. Diatoms consume CO_2 and nutrients to release O_2. There is no waste generation.

1.8. Case study: phycoremediation of Indira Park Lake

1.8.1. Lake location and area

Indira Park is having 76 acres of area and it lies on the lower Tank Bund road downstream of Hussain Sagar, Hyderabad, India. The total lake area is around 1.875 hectare with an average depth of 8 m with a total water volume of 150,000,000 lL approximately with a daily inflow of approximately 500,000 L.

1.8.2. Lake condition before treatment

The lake is heavily contaminated with blue green algae (BGA) mainly *Microcystis* sp. BGA is present as a suspension in the water column all over the lake (**Figure 2**). Inlet water coming from Hussain Sagar Lake is also contributing to BGA input. BGA mat is removed manually on a daily basis. Foul smells emanate from the lake and also from places were lake water is used for horticulture. Lake water contains total nitrogen (TN) and total phosphate (TP) concentration of 35.44 and 1.45 mg/l, respectively COD and BOD were also high at 323 and 64 mg/l, respectively.

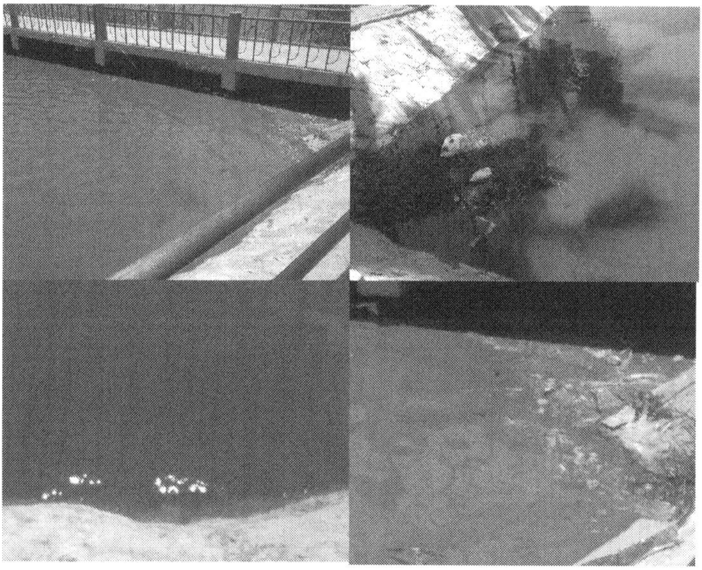

Figure 2. Lake condition before treatment at different locations.

1.8.3. Treatment using Nualgi

Lake treatment was started on May 20, 2014 (Day 01), with the addition of 4 l of Nualgi lakes with a subsequent addition of 2 l on days 6, 14 and 23, respectively. Nualgi was added along the sides of the pond by using a boat, and water samples for testing were collected before addition. Water condition on the day of addition was heavily contaminated with BGA growth and the BGA layer is formed towards the east-side bridge, and as the air turbulence is from west to east, BGA layer was forming on the bridge side (**Figure 2**).

1.8.4. Change in visible water condition and physiochemical parameters

Water quality of the pond changed considerably when we investigate at visible changes of pond surface (**Figure 3**) and also the change in water quality parameters (**Table 1**) tested before and 1 week after treatment a significant reduction in total dissolved solids (TDS) was observed from the initial reading of 864–474 mg/l; similarly, COD and BOD were also reduced from 350 to 212 and 56 to 14 mg/l, respectively. Nutrient levels also reduced with nitrate reducing from 1.94 to 0.86 mg/l and phosphate reducing from 1.12 to 0.88 mg/l and Total Kjeldahl Nitrogen (TKN) also reduced from 16 to 10 mg/l over a period of 7 days after the addition of Nualgi.

Figure 3. Visible change in water colour and BGA layer during treatment with Nualgi.

S. No	Parameter	Inlet	Day 01	Day 06	Day 14	Day 23	Percentage of reduction
	Nualgi dosage		4 l	2 l	2 l	2 l	
1	pH	7.61	6.56	6.98	6.99	7.01	
2	Conductivity, ms/cm	1021	932	856	824	839	
3	TDS, mg/l	935	864	474	948	1079	
4	TSS, mg/l	33	14	<10	<10	<10	
5	COD, mg/l	323	350	212	101	32	94%
6	BOD, mg/l	64	56	14	27	10	89%
7	DO, mg/l	0.3	0.6	0.2	1.2	0.8	
8	TKN, mg/l	33	16	10	18	6	83%
9	Nitrate, mg/l	2.42	1.94	0.86	0.78	0.58	82%
10	Phosphate, mg/l	1.45	1.12	0.88	0.54	0.38	80%
11	Faecal coliform	21	18	8	12	12	

Table 1. Water quality parameters tested with percentage of reduction for inlet water from Hussain Sagar, Indira Park Lake water, during treatment with Nualgi and Nualgi dosage pattern.

1.8.5. Change in phytoplankton diversity

From phytoplankton analysis, BGA was the dominant phytoplankton type present in the lake before Nualgi addition and *Microcystis* sp. was the dominant species along with *Spirulina* sp. but after Nualgi addition slowly the phytoplankton dominance shifted from BGA to diatoms (**Table 2**) with *Nitzschia* sp., *Navicula* sp., *Cocconeis* sp., *Gymphonema* sp. and *Gyrosigma* sp. as dominant species along with *Cyclotella* sp. during the initial phase of treatment. From this analysis, Nualgi clearly triggered diatom growth and in turn diatoms acted as catalysts in the improvement of water quality parameters and reduction of BGA growth.

Phytoplankton	Day 01	Day 07	Day 16	Day 24
Blue green algae (BGA)	122	63	43	41
Diatoms				
Pennate	11	34	54	71
Centric	2	13	16	16

Table 2. Changes in phytoplankton concentration and diversity during lake treatment using Nualgi.

2. Conclusions

Using algae for the reduction of nutrients is a unique technology, which utilizes the enormous potential of microalgae in restoring water quality. Diatoms have the ability to simultaneously tackle more than one problem, which is not capable by conventional chemical processes. Growing microalgae using waste water can provide a viable alternative of tertiary biotreatment coupled with simultaneous production of value-added biomass with various benefits such as production of biofuels, antioxidant, anti-cancerous, anti-obesity, anti-viral, antibacterial compounds, aqua, poultry, animal feed additives, etc. Phycoremediation is cost-effective as it saves power and many chemicals, and it has a potential for CO_2 sequestration—a solution for the threat of global warming.

Author details

Marella Thomas Kiran[1], Mallimadugula Venkata Bhaskar[2] and Archana Tiwari[1*]

*Address all correspondence to: panarchana@gmail.com

1 Department of Biotechnology, Noida International University, Yamuna Expressway, Greater Noida, Uttar Pradesh, India

2 Kadambari Consultants Pvt. Ltd., Ashok Manipuri, Kapra, Hyderabad, India

References

[1] Walker Jr WW. Significance of eutrophication in water supply reservoirs. Journal (American Water Works Association). 1983;75(1):38–42.

[2] Boyd C. Water quality in ponds for aquaculture. Alabama Agriculture Experiment Station Auburn University, Birmingham Publishing Co., Birmingham, AL, 1990.

[3] Codd G. Cyanobacterial toxins: occurrence, properties and biological significance. Water Science and Technology. 1995;32(4):149–56.

[4] Lawton LA, Codd G. Cyanobacterial (blue-green algal) toxins and their significance in UK and European waters. Water and Environment Journal. 1991;5(4):460–5.

[5] Tilman D, Kiesling R, Sterner R, Kilham S, Johnson F. Green, bluegreen and diatom algae: taxonomic differences in competitive ability for phosphorus, silicon and nitrogen. Archiv für Hydrobiologie. 1986;106(4):473–85.

[6] Müller-Navarra DC, Brett MT, Liston AM, Goldman CR. A highly unsaturated fatty acid predicts carbon transfer between primary producers and consumers. Nature. 2000;403(6765):74–7.

[7] Field CB, Behrenfeld MJ, Randerson JT, Falkowski P. Primary production of the biosphere: integrating terrestrial and oceanic components. Science. 1998;281(5374):237–40.

[8] Dugdale RC, Wilkerson FP. Silicate regulation of new production in the equatorial Pacific upwelling. Nature. 1998;391(6664):270–3.

[9] Craggs RJ, Adey WH, Jenson KR, John MSS, Green FB, Oswald WJ. Phosphorus removal from wastewater using an algal turf scrubber. Water Science and Technology. 1996;33(7):191–8.

[10] Congestri R, Cox EJ, Cavacini P, Albertano P. Diatoms (Bacillariophyta) in phototrophic biofilms colonising an Italian wastewater treatment plant. Diatom Research. 2005;20(2): 241–55.

[11] Guzzon A, Bohn A, Diociaiuti M, Albertano P. Cultured phototrophic biofilms for phosphorus removal in wastewater treatment. Water Research. 2008;42(16):4357–67.

[12] Olguí EJ. Phycoremediation: key issues for cost-effective nutrient removal processes. Biotechnology Advances. 2003;22(1):81–91.

[13] Oswald WJ, Gotaas HB. Photosynthesis in sewage treatment. Transactions of the American Society of Civil Engineers. 1957;122(1):73–97.

[14] Oswald WJ. Large-scale algal culture systems (engineering aspects). Micro-algal biotechnology. Cambridge University Press, Cambridge, UK. 1988, pp. 357–94.

[15] Oswald WJ, Gotaas HB. Photosynthesis in sewage treatment. Transactions of the American Society of Civil Engineers. 1957;122:73–105.

[16] Oswald WJ. Advanced integrated wastewater pond systems. Proceedings of the ASCE Convention: Supplying Water and Saving the Environment for Six Billion People; EE Div/ASCE, San Frisco, CA. 1990: 73–80.

[17] Horner R. A Taxonomic guide to some common marine phytoplankton biopress. Bristol, England, UK. 2002, pp. 1–195.

[18] Walsh JJ, Dieterle DA, Meyers MB. A simulation analysis of the fate of phytoplankton within the Mid-Atlantic Bight. Continental Shelf Research. 1988;8(5–7):757–87.

[19] Mann DG. The species concept in diatoms. Phycologia. 1999;38(6):437–95.

[20] Hasle G, Syvertsen E. Marine diatoms. Identifying marine phytoplankton. Academic Press, San Diego, CA. 1997, pp. 5–385.

[21] Furnas MJ. In situ growth rates of marine phytoplankton: approaches to measurement, community and species growth rates. Journal of Plankton Research. 1990;12(6):1117–51.

[22] Smetacek V. Diatoms and the ocean carbon cycle. Protist. 1999;150(1):25–32.

[23] Amano Y, Takahashi K, Machida M. Competition between the cyanobacterium *Microcystis aeruginosa* and the diatom *Cyclotella* sp. under nitrogen-limited condition caused by dilution in eutrophic lake. Journal of Applied Phycology. 2012;24(4):965–71.

[24] Treguer P, Nelson DM, Van Bennekom AJ, DeMaster DJ. The silica balance in the world ocean: a reestimate. Science. 1995;268(5209):375.

[25] Ryther JH, Officer CB. Impact of nutrient enrichment on water uses. Estuaries and nutrients. Springer, New York, NY. 1981, pp. 247–61.

[26] Officer C, Ryther J. The possible importance of silicon in marine eutrophication. Marine Ecology Progress Series. 1980;3(1):83–91.

[27] Schelske CL, Stoermer EF, Fahnenstiel GL, Haibach M. Phosphorus enrichment, silica utilization, and biogeochemical silica depletion in the Great Lakes. Canadian Journal of Fisheries and Aquatic Sciences. 1986;43(2):407–15.

[28] Sommer U. The paradox of the plankton: fluctuations of phosphorus availability maintain diversity of phytoplankton in flow-through cultures. Limnology and Oceanography. 1984;29(3):633–6.

[29] Sommer U, Kilham SS. Phytoplankton natural community experiments: a reinterpretation. Limnology and Oceanography. 1985;30:436–40.

[30] Hecky R, Kilham P. Nutrient limitation of phytoplankton in freshwater and marine environments: a review of recent evidence on the effects of enrichment. Limnology and Oceanography. 1988;33(4):796–822.

[31] Suman K, Kiran T, Devi UK, Sarma NS. Culture medium optimization and lipid profiling of Cylindrotheca, a lipid-and polyunsaturated fatty acid-rich pennate diatom and potential source of eicosapentaenoic acid. Botanica marina, 55 (2012) 289–299.

[32] Kiran T, Tiwari A, Bhaskar MV. A new novel solution to grow diatom algae in large natural water bodies and its impact on CO_2 capture and nutrient removal. Journal of Algal Biomass Utilization. 2015;6(2):22–7.

Section 3

Water Quality, Management and Modeling

Chapter 7

Adaptive Management of an Imperiled Catostomid in Lake Mohave, Lower Colorado River, USA

Brian R. Kesner, Jamie B. Wisenall and Paul C. Marsh

Additional information is available at the end of the chapter

http://dx.doi.org/10.5772/63808

Abstract

Lake Mohave, a man-made reservoir in the lower Colorado River, USA, was once home to the largest wild population of the endemic and endangered razorback sucker *Xyrauchen texanus*, estimated at 60,000 individuals in the late 1980s. Individuals of this population were 25 years or older because recruitment was precluded by the removal of larval production by introduced centrarchid species. A repatriation program was initiated in the 1990s to replace the aging population with young fish by capturing larvae from the reservoir and raising them in hatcheries and protected lakeside backwaters until they were released back into the reservoir. Although more than 200,000 fish have been repatriated to Lake Mohave, the repatriate population has remained at a few thousand fish. The wild population is now functionally extinct. The program has adapted to new threats to the population, political realities, and technological advances. Management shifted in 2006 to the Lower Colorado River Multi-Species Conservation Program, which has politicized the process. The aim of this chapter is to describe the initial, informal adaptive management strategy for razorback sucker in Lake Mohave, the transition to a formal program, and the inherent pitfalls that formalization entails.

Keywords: endangered species, population dynamics, genetic diversity, razorback sucker, hatchery management

1. Introduction

Razorback sucker *Xyrauchen texanus* is a long-lived catostomid that is endemic to the Colorado River basin of western North America. Historically, it was encountered throughout the basin in the mainstem Colorado River as well as medium to large tributaries [1]. Declines in abundance and range contraction over the last half century are attributed to habitat alteration (e.g., dams and water extraction) and introduction and establishment of more than four dozen nonnative fish species [2]. Razorback sucker was federally (USA) listed as endangered in 1991 [3], and management actions divided the basin into upper and lower units at Lee's Ferry upstream of the Grand Canyon (**Figure 1**). In the lower Colorado River, management actions were initially focused on the largest remnant population of razorback sucker in Lake Mohave, a water regulation reservoir impounded in 1951 by the completion of Davis Dam (**Figure 2**).

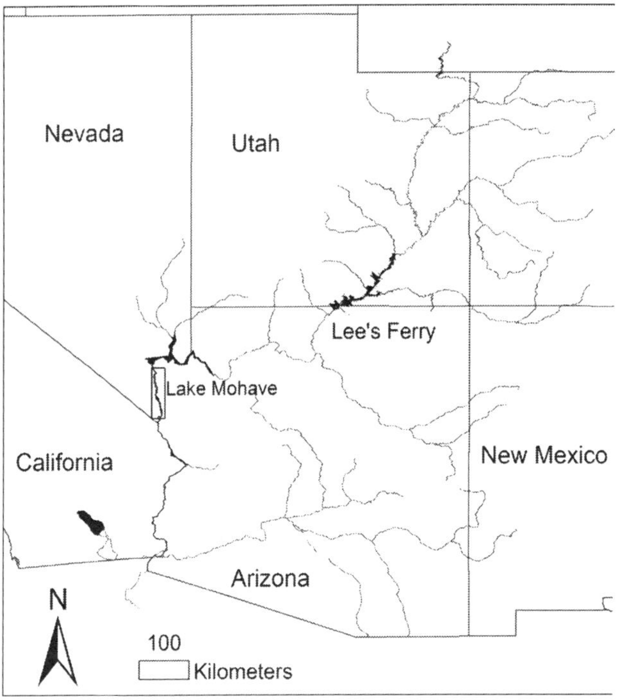

Figure 1. Map of the Colorado River drainage in western USA.

The management of razorback sucker in Lake Mohave began with a single goal, to replace a declining old population likely to become extirpated by the turn of the twentieth century with a young repatriated population [4]. This single goal was established in the early 1990s by the ad hoc Lake Mohave Native Fishes Workgroup (NFWG), an informal group of biologists from state and federal agencies, universities, and private entities. It was seen as a stop-gap measure to buy time until effective, long-term recovery actions could be developed and implemented. Nearly 30 years later, the management strategy has resulted in a genetically diverse repatriate population of approximately 2500 individuals [5], far fewer than the estimated 60,000 wild adults that resided in the lake in the late 1980s. However, without this repatriated population, razorback sucker would have disappeared from the reservoir, as the wild population is now functionally extirpated [5].

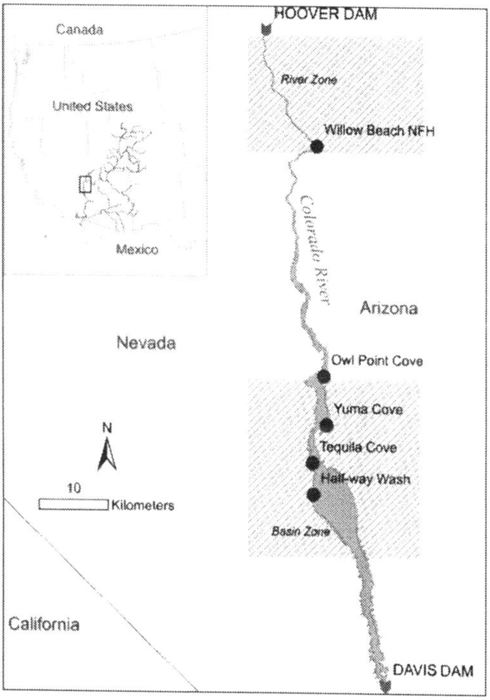

Figure 2. Map of Lake Mohave Arizona and Nevada, USA.

The management strategy for razorback sucker in Lake Mohave has always been adaptive. As is likely with any resource management program that has covered multiple decades, it had to adapt to changing environmental conditions, technological advancements, and periodic political intransigence. Today, adaptive management is at the cornerstone of federal policy regarding endangered species recovery plans [6]. Recently, the informal adaptive management strategy for razorback sucker in Lake Mohave (NFWG) has been replaced by the formal and well-documented adaptive management structure of the Lower Colorado River Multi-Species Conservation Plan (LCR MSCP). This plan seeks to conserve at least 26 plant and animal species in the lower Colorado River corridor while continuing to allow normal Colorado River water and power operations [7]. Applying adaptive management principles to endangered species recovery programs requires striking a delicate balance. The public and elected officials often demand an end date, an exact price in dollars, and time for a recovery program, which can run counter to the adaptive management approach because new data can change predicted outcomes of management practices.

2. Adaptive management of razorback sucker in Lake Mohave

2.1. Population decline and razorback sucker repatriation program

The number of razorback sucker in Lake Mohave was likely in the hundreds of thousands, if not millions in the 1960s and 1970s, but no attempt to estimate their abundance was made until the early 1980s [8]. By the late 1980s, the remnant population was estimated at 60,000 individuals [9]. This was the largest remaining population of razorback sucker in existence, but fish that had been aged through otolith examination were old, between 25 and 45 years old [10]. Most of these adults were representatives of successful cohorts produced in the first few years after impoundment of the reservoir, and no recruitment had been detected for 25 years. If this continued, the population was projected to disappear before the turn of the century [1, 11].

Although the immediate threat to the species within Lake Mohave was simple to identify, no simple solution was available. The complete removal of larval production from the entire reservoir through predation, mainly from introduced centrarchid species, was seemingly improbable, but multiple studies spanning several decades throughout the basin all pointed in the same direction [12, 13]. Data to support alternative hypotheses could not be produced. Resource limitation appeared promising and was carefully investigated but could not be clearly identified as a mortality factor [14, 15]; larval razorback sucker survived and grew inside protective enclosures placed in the reservoir, and razorback sucker grew to adults when placed in a nonnative free environment. Neither the political climate nor the technical knowhow was available at the time or is presently available to remove all nonnative fish species from the reservoir. Therefore, if the population and the species were to be perpetuated, assistance would have to take the form of stocking fish.

Early attempts to produce juveniles took place in an isolated backwater adjacent to Lake Mohave at Yuma Cove where a large aggregation of fish was available during the spawning season [1]. The first attempt in January 1991 was to stock ripe fish captured from the reservoir (33 females and 67 males) directly into the backwater. Larvae were produced but no juveniles survived; at least none was detected, and no mortality factor was identified. A year later in January 1992, 28 females and 60 males were transferred from the lake into the backwater. Larvae again were produced and this time juveniles survived and nearly 300 were captured the following autumn. However, genetic evaluation indicated that the juveniles represented a relatively small number of parents, their variation was less than expected relative to the wild adult population, and this method of propagation was unlikely to preserve the population's genetic diversity [5, 16, 17]. The next iteration in March 1993 involved manually spawning ripe fish on site and stocking about 200,000 embryos thus produced into the backwater. Unfortunately, water level in the backwater lowered unexpectedly and exposed the bottom where most fertilized ova had settled, killing them. Nonetheless, some larvae hatched and a small number of juveniles was captured. At the same time, a number of laboratory-reared fish (metalarvae and juveniles averaging 26 mm long) were stocked into this and other backwaters around the lake. Survival of these fish was variable among sites, but recovery of fewer than 500 juveniles was inadequate to fulfil programmatic goals of stocking thousands of young fish.

http://dx.doi.org/10.5772/63808

The real "aha" moment came with the suggestion to harvest naturally produced larvae directly from the lake and transfer them to protective custody in nonnative-free rearing sites including lakeside backwaters. Larvae were known to be phototactic [8] and easily captured, abundant, and likely to represent the genetic diversity of the wild adult population. Thus, in winter-spring 1994 began the methodical process of nighttime harvest with dip nets of larvae that were attracted to lights, rearing in safe places, and return to the lake as relatively large subadults or adults. One unique aspect of the program was the early use of passive-integrated transponder (PIT) tags. From the beginning, every attempt was made to implant all repatriated razorback sucker with a PIT tag prior to being released into the reservoir. Each tag contained a unique 10-digit hexadecimal code that was used to identify individual fish. With periodic refinements, the same basic protocol is followed today.

Early indications were that the repatriation program was a success. By 1994, repatriated razorback sucker were captured in routine monitoring on the spawning grounds throughout the reservoir. To increase the capacity of the program, the NFWG partnered with golf courses in Boulder City, Nevada, Willow Beach National Fish Hatchery (a federal trout hatchery that was built on the shores of the reservoir downstream of Hoover Dam), and other entities. An average of more than 12,000 razorback sucker was repatriated annually from 1997 to 2005. In 1999, the repatriate population was estimated at more than 1000 fish, based on PIT tag recapture data [18]. However, from 1999 through 2004 the repatriate population appeared to plateau at about 1500 fish (estimates fluctuated between 1000 and 2500) regardless of the number stocked (**Figure 3**). It was clear that the repatriate population was not going to match the previous size of the wild population under the current program. Meanwhile, the wild population had declined from more than 60,000 in 1991 to less than 50 [5], and now is functionally extinct. Concerns over whether genetic diversity could be maintained long term with a population of only a few thousand [19] motivated research into the sources of post-stocking mortality.

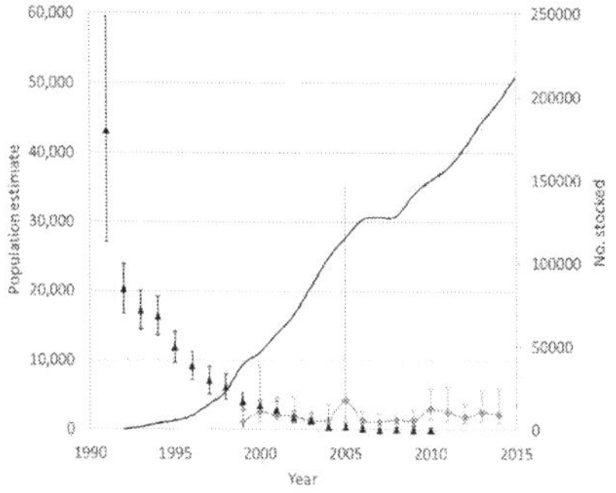

Figure 3. Wild (black triangles) and repatriate (gray diamonds) population estimates and 95% confidence intervals for razorback sucker in Lake Mohave. Estimates are single census mark-recapture estimates derived from captures during annual netting activities in March. The solid line and secondary axes represent the cumulative number of PIT tagged repatriated razorback sucker stocked into Lake Mohave.

Based largely on work in the Gila River, the program originally considered a 250-mm razorback sucker to be free from most predatory threats in Lake Mohave, as the main concerns were channel catfish *Ictalurus punctatus* and centrarchid species common to the reservoir, for example, largemouth bass *Micropterus salmoides* and bluegill *Lepomis macrochirus*. An unexpected population expansion of one of the reservoir's nonnative predatory fishes altered this condition at nearly the same time the repatriation program was initiated. During a season of heavy winter rains in 1983, the spillway was opened on Hoover Dam. Besides releasing water downstream unfettered, this operation allowed the introduction into Lake Mohave of large numbers of the obligate piscivore striped bass *Morone saxatilis*. Although this species already was known to occur in Lake Mohave [20], the influx of individuals and their subsequent reproduction and recruitment led to it becoming a predominant species by 1990. The impact of striped bass was immediate and severe, yet interestingly the phenomenon is largely unstudied except in the context of native fishes. The trout fishery downstream of Hoover Dam was considered to be one of the best in the United States in the 1980s [21], but trout were rarely seen more than a few weeks after stocking by the mid-1990s. Striped bass grew rapidly in Lake Mohave obtaining sizes of up to 1200 mm TL and more than 30 kg in weight. The target-stocking size of 250 mm for razorback sucker was clearly inadequate, but the "right" size to mitigate striped bass predation was unknown.

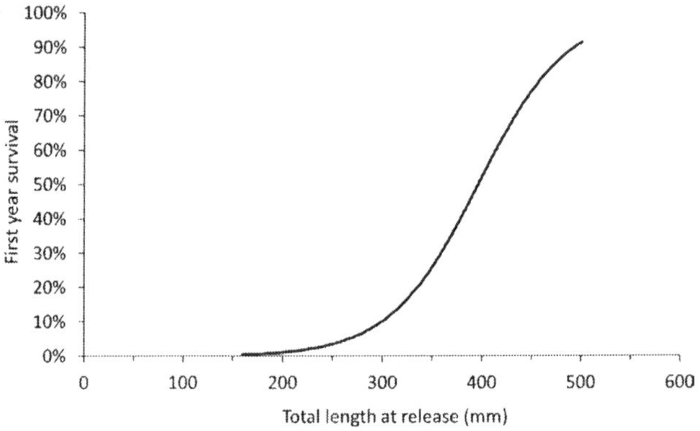

Figure 4. Estimated first-year survival for repatriated razorback sucker released into Lake Mohave based on total length (TL) when released. Survival estimates based on mark-recapture analysis of release and capture data in Lake Mohave from 1993 through 2004.

One of the early attempts to quantify the size-survival relationship for razorback sucker used PIT tag recapture data from annual census data. Sampling trips to monitor populations of native (and endangered) bonytail *Gila elegans* and razorback sucker in Lake Mohave were conducted multiple times per year, but most razorback sucker were captured on sampling trips conducted during the peak of spawning activity in March. These "March Roundup" recapture data were linked to stocking records to derive capture histories for all repatriated fish released in Lake Mohave. TL at release was added as a covariate to first-year survival and the data were assessed within the mark-recapture software MARK. The resulting relationship between first-year survival and TL at release was used as a management tool (**Figure 4**), but there was considerable uncertainty about the relationship due to low survival of stocked fish, less than 2% overall, and low annual recapture rates, less than 10% [18]. Increasing recapture rate would require increasing netting effort during the March Roundup. Members of the NFWG were concerned about increasing stress and potential delayed mortality due to repeated handling of fish. The program needed to adapt to increase survival of stocked fish, and increase recapture rates without increasing handling.

2.2. Research, adaptation, and emerging technology

Prior to the formal publication of the size-survival relationship, incremental increases in stocking sizes were implemented moving to 300 and 350 mm TL from 1999 to 2004. By 2006, the decision was made to produce fish that were nearly immune to predatory threats, which required raising razorback sucker over 500 mm TL. Raising large numbers of razorback sucker to 500 mm TL at the hatchery was going to take additional years of growth and experimentation. In the interim acoustic telemetry studies were performed with small numbers of razorback sucker at the potentially optimal stocking size of 500 mm TL and at the previously common release size of 380 mm TL to test the hypothesis. The results from these studies were unequivocal and confirmed the size-survival relationship, but survival also varied dramatically from year to year [22]. In addition, an acoustic tag that was implanted in a 520-mm TL razorback sucker was recovered from the stomach of a relatively average sized 13-kg striped bass by a fisherman [23], providing evidence that no size was completely safe from predation; the maximum length of razorback sucker is thought to be near a meter [24], but few individuals longer than 700 mm now are encountered [18, 20].

While razorback sucker were being held for additional years of growth at the hatchery, few fish were stocked; only about 500 repatriates were released from the hatcheries between November 2007 and October 2009. The 2007 population estimate, calculated after the 2008 March Roundup, had dropped to near 1000 adult repatriates. Although razorback sucker held at the hatcheries had not reached the target length of 500 mm TL, the NFWG determined that releasing fish at their current size was better than taking the risk that genetic diversity of the population would be compromised due to the low population size. Over the next 18 months, more than 20,000 razorback sucker were stocked from the hatchery system. Although the mean TL of all fish stocked was similar to previous stockings (360 mm), more than 800 razorback sucker at 450 mm TL or longer were released.

Along with a shift in release size, a technological shift also occurred in 2006. PIT tagging all repatriates released into Lake Mohave had been conducted since the program began in 1992 with few exceptions. By 2006, PIT tag technology had progressed significantly. The 400-kHz tags used in Lake Mohave for the previous 14 years were being superseded by 134.2-kHz PIT tags. Some PIT tag readers could detect both tag frequencies, but the newer 134.2-kHz tags were becoming the dominant format in animal tagging. The 134.2-kHz tag had significant advantages over the older frequency. The 400-kHz tags were read by rubbing the PIT tag reader over the tag injection site, the new 134.2-kHz tags could be read from tens of centimeters away.

Not only did this reduce handling time by making it easier during capture to identify fish, it was now possible to detect fish without handling them. After years of favoring methodological continuity over emerging technology, the shift was made to 134.2-kHz tags in 2006. All razorback sucker were now tagged with the new tag prior to repatriation, and any razorback sucker captured during monitoring activities without a tag was given the new tag. Due to the lack of stocking from 2006 through 2008, razorback sucker captured with 400-kHz tags were given a 134.2-kHz tag starting in 2008 (double tagging).

Oversight of the management program also shifted by 2006. Prior to 2006, management of the program was guided by the NFWG. However, in 2005 the Lower Colorado River Multi-Species Conservation Plan was signed. This negotiated agreement between state water resource agencies of the lower Colorado River basin and the federal government is a 50-year agreement for continued operations and conservation actions for at least 26 native plant and animal species. The informal adaptive management approach of the NFWG was replaced with a formal adaptive management strategy when the LCR-MSCP was signed and became the de facto management doctrine for razorback sucker and other native fish species in the lower Colorado River. The LCR MSCP strategy for the conservation of razorback sucker in the lower Colorado River closely followed the concepts developed by Minckley et al. [25]. This paper acknowledged the limited success of the repatriation program, and advocated for a different approach to conserving razorback sucker long term: create self-recruiting populations in off-channel habitats free of nonnative fishes, while maintaining a genetically diverse population of adults in the mainstem Colorado River and its reservoirs. The Lake Mohave razorback sucker population continued to be important as the genetic repository of the species, and understanding the population dynamics of the off-channel and reservoir populations would be fundamental to the success of the program. Remote PIT scanning appeared to be the tool needed to acquire the data upon which to base science-driven management decisions.

The increased reception range of the new 134.2-kHz PIT tag allowed for remote sensing of PIT tags, that is, identifying a tagged fish without capturing it. Elsewhere, portable PIT scanners were used to monitor behavior, movement, and habitat use of fishes in shallow waters of small streams [26, 27], and fish movement has also been monitored in larger streams using units permanently or semi-permanently mounted to the substrate or man-made structure [28, 29]. Initial testing of this technology in Lake Mohave relied on Biomark® flat-plate antennas and FS2001 radiofrequency identification (RFID) scanners. These units were able to detect both 400-kHz and 134.2-kHz tags. Although read range of 400-kHz PIT tags was only a few centimeters or less, razorback sucker were known to remain close to the substrate and the potential to remotely contact at least a portion of 400-kHz tagged fish was important given the small number of 134.2-kHz tagged razorback sucker released by 2008. These PIT scanning units could only be deployed in the calm, lotic waters of the reservoir because they required a floating component to contain the RFID readers and batteries. In 2008, the remote PIT-scanning units on Lake Mohave contacted 176 unique PIT tags and 194 were contacted in 2009 [30]. In 2010, the number of available 134.2-kHz tagged fish increased due to large stocking events. PIT scanning in the reservoir contacted 477 razorback sucker in that year, nearly twice as many as the number encountered during the March Roundup (286 razorback sucker) in the same

http://dx.doi.org/10.5772/63808

year. It was clear that this technology would allow for a much larger proportion of the population to be contacted annually than the netting effort once the 134.2-kHz tagged fish made up the majority of the population. This increase in monitoring data also would not require additional handling or netting activities, which could impact survival or spawning behavior. However, the PIT scanners could not alleviate one monitoring issue. Razorback sucker were routinely observed in the swiftly flowing and turbulent riverine section of the reservoir downstream of Hoover Dam, but these fish were generally unmonitored. Trammel nets were ineffective in the fast-flowing river, and boat electrofishing was effective, but potentially harmful to the fish [31]. PIT scanning would likely be effective as well, but the floating systems could not be anchored well enough to remain in place in the river.

In 2010, a PIT scanner design was proposed for use in the riverine sections of Lake Mohave based on systems described in Bond et al. [28]. The design was adapted to allow the unit to be completely self-contained without a floating or a shore-based component. The first fully submersible PIT scanners were developed and deployed in 2011 between Hoover Dam and Willow Beach. Sites of deployment were based on observations of spawning aggregates during the winter and spring months (January through April), and PIT scanner units were typically deployed between 12 and 24 h before being retrieved, downloaded, and redeployed with fresh batteries during week-long sampling trips. In that first year, 670 PIT tags from repatriated razorback sucker were contacted. By 2012, this number nearly tripled to 1832, more than twice the number contacted in the basin by PIT scanners (882). The total number scanned in the river alone nearly matched the 2012 population estimate of 1854. Combined, repatriate PIT scanning contacts in Lake Mohave exceeded the population estimate based on mark-recapture data from netting activities during the March Roundup.

One purpose of extending PIT scanning to the riverine section below Hoover Dam was to determine if the fish in the river and in the basin were of the same population. It was assumed that razorback sucker moved throughout the reach between Davis and Hoover dams, and that population estimates based on netting activities on the spawning grounds in the basin were representative of the lake-wide population. Initial analysis of PIT scanning data from 2011 and 2012 in the basin and river did not support this assumption. More than 80% of the fish contacted in the riverine reach downstream of Hoover Dam had been released (stocked) there, and more than 80% of the fish contacted in the basin had been released in the basin. In year-to-year comparisons, fewer than 10% of the fish contacted in one reach were contacted in the other reach the subsequent year. The partial demographic isolation of the two subpopulations required a rethinking of population estimates and the overall stocking strategy.

Beginning in 2012, year-to-year population estimates based on PIT scanning data supplemented the annual estimates based on annual netting activities. Consistent with the partial demographic isolation demonstrated in the PIT scanning data, population estimates were divided into basin and river subpopulations. In the first few years, the sum of subpopulation estimates based on PIT scanning was similar to estimates from netting activities even though the river subpopulation was not effectively sampled by netting activities. This was likely due to the exclusion of 400-kHz tagged fish in PIT scanning and the lack of complete geographic coverage of the basin area with PIT scanners. Netting activities generally covered four primary spawning aggregates in the basin area: Halfway Wash, Tequila Cove, Yuma Cove, and Owl Point. The majority of PIT scanning in 2011 and 2012 was concentrated in Tequila Cove and Yuma Cove. These two sites had semi-permanent PIT scanners that scanned nearly continuously throughout the spawning season (November through April). As the geographic coverage in the basin improved, and the proportion of 134.2-kHz tagged fish increased, the subpopulation estimate for the basin alone approached the overall estimate based on netting activities. By 2014, each subpopulation was estimated at around 1500 fish each. The discovery of additional razorback sucker in the system was a positive development for the program. However, the river subpopulation did not contribute significantly to the repatriation program. This is because until recently all larvae collected in the lake for the repatriation program were collected more than 10 km downstream of Willow Beach, whereas the majority of the river subpopulation was reproducing upstream reproducing upstream of Willow Beach. Proportional representation of the riverine subpopulation in larval collections would require about half of the 20,000 larvae collected annually to be collected upstream of Willow Beach, and so the program must again adapt to this new paradigm.

3. LCR MSCP and the future of Lake Mohave

The original goal of replacing the wild adult population with a young repatriated population was based on solid scientific evidence available at the time it was formulated. Changes in the fish fauna and lower than expected adult survival resulted in a population of only a few thousand, maintained through annual stocking of more than 10,000 razorback sucker. The NFWG adapted to new data and technologies, and the NFWG recognized the futility of continuing the stocking program indefinitely although a small population of razorback sucker had been established and maintained. Alternative strategies to the repatriation program were under development as early as the late 1990s. The overall strategy for razorback sucker conservation in the lower Colorado River codified in journal publications [5, 25], the US Fish and Wildlife Service (USFWS) implementation plans [32], and the LCR MSCP is one of natural recruitment within protected off-channel habitats. These small populations would be managed with the exchange of adults with the populations maintained in the mainstem and its reservoirs and with each other to maintain genetic diversity. This would eliminate the need for hatcheries and massive stockings and permit nearly natural selective pressures to continue. The LCR MSCP conceptually is in agreement with this strategy, but its formal adaptive management structure spreads limited resources too thin and its political nature creates obstacles to implementation of new strategies when data are acquired.

The basic concept of adaptive management is to treat current management practices as working hypotheses, and the results of such practices are evaluated through monitoring to provide information, which forms the basis of changes in subsequent management practices. The number of knowledge gaps identified by the LCR MSCP conceptual model for razorback sucker [33] that required additional research stretches scant resources to their limit, and does nothing for conserving the species in the short term. The immediate need of optimizing stocking regimes should take precedence over understanding the complete life cycle of the species, especially when most of that research requires experiments with life stages that cannot be found in the natural system. Razorback sucker as a species is still in crisis, the size of the one population with the genetic legacy is too low to secure that legacy. Does it therefore make sense to spend resources answering biological questions that at best will aid razorback sucker recovery which is decades off and likely will not happen at all if current trends continue?

One example of the lack of focus since the LCR MSCP is the size at release for razorback sucker in Lake Mohave. After the release of large fish from the hatchery system from 2009 to 2011, an experiment that began prior to the LCR MSCP, the target released size returned to 300 mm TL. The only large fish released into Lake Mohave have been from lakeside backwaters. This has resulted in almost no razorback sucker stocked in the last 3 years from the hatchery being captured or contacted via PIT scanners. For comparison, out of the 806 fish longer than 450 mm TL that were released during the attempt to grow 500-mm TL fish, 417 were contacted by PIT scanners from 2011 through 2015. More than 22,000 razorback sucker were released in 2012 and 2013 (41 fish over 450 mm TL), and only 169 have been contacted more than 30 days after their release. The complete failure of recently stocked fish to recruit to the adult population has erased any gains in population size that were created from the large fish releases in 2009, 2010, and 2011.

Equally important, the political constraints cause actions that could benefit razorback sucker to not be performed. The LCR MSCP mandates the creation of 360 surface acres of backwater. The most successful backwaters in the system to date are the lakeside backwaters on Lake Mohave. Experiments designed to assess the genetic contribution of one razorback sucker generation to the next in backwater environments are currently being conducted in three lakeside backwaters on Lake Mohave. Two of these backwaters produce young of year razorback sucker annually, but must be harvested each year as the backwaters dry up during reservoir drawdowns. One lakeside backwater at Yuma Cove that is deep enough to retain water year-round has a healthy population of several hundred razorback sucker, some stocked, some self-recruited in the backwater. With the exception of Cibola High Levee Pond [25], there is currently no other location on the lower Colorado River that has proven as effective. However, the LCR MSCP does not give credit to the program for any off-channel habitat created in Lake Mohave, and unfortunately the lakeside backwaters will only be used for experiments or grow-out, and no new backwaters will be created on the reservoir.

4. Conclusions

The real danger of an expensive management program for any endangered species is failure to produce tangible results. A perception that millions of tax dollars were spent without a return on the investment would likely reduce support for future programs. The endangered fish recovery program in the upper basin should be a warning sign for the LCR MSCP. Although the LCR MSCP is very specifically a conservation and not a recovery program, that distinction will likely not save it from public or political scrutiny if it is perceived a failure. A 30- or 50-year time span may seem like a long time to recover a species, but it is within the life span of a single razorback sucker. It is unreasonable to expect that our understanding of the causes of population decline for razorback sucker will reach the level of being able to reverse that decline within the life span of a single fish. The upper basin has had nearly three decades to reach that goal, yet not one population of razorback sucker in that region is doing better now than before the beginning of the program if impacts of stocking are excluded. Although stocking can reduce the probability of extinction, it does nothing to improve the probability of self-sustenance, which is a fundamental requirement of any successful recovery program.

The ultimate fate of the razorback sucker population in Lake Mohave under the LCR MSCP is unknown, but if the NFWG had required that a complete understanding of razorback sucker life history and predator-prey interactions be obtained prior to stocking one fish, there would be no population to conserve. The NFWG was able to focus on a single goal, adapted to new information, and still failed to achieve its goal. The LCR MSCP is attempting to achieve multiple goals, with a 5-year planning cycle, a 54-member steering committee, and an expectation that "program-level adaptive management is not anticipated to occur often over the 50 year term because the HCP conservation direction was developed using the best available information" [34]. There recently has been a refocus on meeting a target release size of near 500 mm TL. If razorback sucker at or near 500 mm TL are released consistently, a substantial increase in the number of individuals in the population is expected. These actions will secure the razorback sucker in Lake Mohave for a number of years; however, conservation and recovery of the species in the lower Colorado River will require successful implementation of the backwater-based program concept [25, 32]. Such a plan already has proven its biological efficacy and we urge the LCR MSCP partners to move forward aggressively to identify, confront, and overcome political and logistical barriers and put that plan into practice.

Author details

Brian R. Kesner*, Jamie B. Wisenall and Paul C. Marsh

*Address all correspondence to: bkesner@nativefishlab.net

Marsh & Associates LLC, Tempe, AZ, USA

References

[1] Minckley WL, Marsh PC, Brooks JE, Johnson JE, Jensen BL. Management toward recovery of the razorback sucker. In: Minckley WL, Deacon JE, editors. Battle Against Extinction: Native Fish Management in the American West. Tucson, AZ, USA: University of Arizona Press; 1991. p. 303–317.

[2] Minckley WL. Native fishes of the Grand Canyon region: an obituary? In: Colorado River Ecology and Dam Management; 24–25 May 1990; Santa Fe, NM. Washington DC: National Academy Press; 1991. p. 124–177. DOI: 10.17226/1832

[3] United States Fish and Wildlife Service. Endangered and threatened wildlife and plants; the razorback sucker (*Xyrauchen texanus*) determined to be an endangered species, final rule. Federal Register. 1991;56:54597–54967.

[4] Mueller GA. A program for maintaining the razorback sucker in Lake Mohave. Bethesde, MD. In: American Fisheries Society Symposium 15; American Fisheries Society; 1995. p. 127–135.

[5] Marsh PC, Dowling TE, Kesner BR, Turner TF, Minckley WL. Conservation to stem extinction: the fight to save razorback sucker *Xyrauchen texanus* in Lake Mohave and its implications for species recovery. Copeia. 2015;103:141–156.

[6] Doremus H. Adaptive management, the Endangered Species Act, and the institutional challenges of new age environmental protection. Washburn Law Journal. 2001;41(1): 50–89.

[7] Lower Colorado River Multi-Species Conservation Program, Volume II: Habitat Conservation Plan. Sacramento, CA; 2004. 506 p.

[8] Bozek MA, Paulson LJ, Deacon JE. Factors affecting reproductive success of bonytail chubs and razorback suckers in Lake Mohave. Report to US Fish and Wildlife Service Contract No. 14-16-0002-81-251. Las Vegas: University of Nevada; 1984. 136 p.

[9] Marsh PC, Minckley WL. Status of bonytail (*Gila elegans*) and razorback sucker (*Xyrauchen texanus*) in Lake Mohave, Arizona-Nevada. Proceedings of the Desert Fishes Council. 1992;23:18–23.

[10] McCarthy MS, Minckley WL. Age estimation for razorback sucker (Pisces: Catostomidae) from Lake Mohave, Arizona and Nevada. Journal of the Arizona-Nevada Academy of Science. 1987;21:87–97.

[11] Marsh PC, Pacey CA, Kesner BR. Decline of the razorback sucker in Lake Mohave, Colorado River, Arizona and Nevada. Transactions of the American Fisheries Society. 2003;132:1251–1256.

[12] Marsh PC, Langhorst DR. Feeding and fate of wild larval razorback sucker. Environmental Biology of Fishes. 1985;21:59–67.

[13] Carpenter J, Mueller GA. Small nonnative fishes as predators of larval razorback suckers. The Southwestern Naturalist. 2008;53:236–242.

[14] Papoulias D, Minckley WL. Food limited survival of larval razorback sucker, *Xyrauchen texanus*, in the laboratory. Environmental Biology of Fishes. 1990;29:73–78.

[15] Horn MJ. Nutritional limitation of recruitment in the razorback sucker (*Xyrauchen texanus*) [dissertation]. Tempe: Arizona State University; 1996. 280 p.

[16] Dowling TE, Minckley WL, Marsh PC. Mitochondrial DNA diversity within and among populations of razorback sucker (*Xyrauchen texanus*) as determined by restriction endonuclease analysis. Copeia. 1996;(3):542–550.

[17] Dowling TE, Minckley WL, Marsh PC, Goldstein E. Mitochondrial DNA diversity in the endangered razorback sucker (*Xyrauchen texanus*): analysis of hatchery stocks and implications for captive propagation. Conservation Biology. 1996;10:120–127.

[18] Marsh PC, Kesner BR, Pacey CA. Repatriation as a management strategy to conserve a critically imperilled fish species. North American Journal of Fisheries Management. 2005;25:547–556.

[19] Dowling TE, Marsh PC, Kelsen AT, Tibbets CA. Genetic monitoring of wild and repatriated populations of endangered razorback sucker. Molecular Ecology. 2005;14(1):123–135.

[20] Minckley WL. Status of razorback sucker, *Xyrauchen texanus* (Abbot), in the lower Colorado River basin. The Southwestern Naturalist. 1983;28:165–187.

[21] Allan RC, Roden DL. Fish of Lake Mead and Lake Mohave. Reno, Nevada: Nevada Department of Wildlife Biological Bulletin No. 7; 1978. 105 p.

[22] Karam AP, Kesner BR, Marsh PC. Acoustic telemetry to assess post-stocking dispersal and mortality of razorback sucker *Xyrauchen texanus*. Journal of Fish Biology. 2008;73:719–727.

[23] Karam AP, Marsh PC. Predation of adult razorback sucker and bonytail by striped bass in Lake Mohave, Arizona-Nevada. Western North American Naturalist. 2010;70:117–120.

[24] Minckley WL. Fishes of Arizona. Phoenix, AZ: Arizona Game and Fish Department; 1973. 293 p.

[25] Minckley WL, Marsh PC, Deacon JE, Dowling TE, Hedrick PW, Matthews WJ, Mueller G. A conservation plan for native fishes of the lower Colorado River. Bioscience. 2003;53:219–234.

[26] Riley WD, Eagle MO, Ives MJ, Rycroft P, Wilkinson A. A portable passive integrated transponder multi-point decoder system for monitoring habitat use and behaviour of freshwater fish in small streams. Fisheries Management and Ecology. 2003;10:265–268.

[27] Roussel JM, Cunjak RA, Newbury R, Caissie D, Haro A. Movement and habitat use by PIT-tagged Atlantic salmon parr in early winter: the influence of anchor ice. Freshwater Biology. 2004;49:1026–1035.

[28] Bond MH, Hanson CV, Hayes SA, Baertsch R, Mcfarlane RB. A new low cost in-stream antenna for tracking passive integrated transponder (PIT) tagged fish in small streams. Transactions of the American Fisheries Society. 2007;136:562–566.

[29] Lucas MC, Mercer T, Armstrong JD, McGinty S, Rycroft P. Use of a flat-bed passive integrated transponder antenna array to study the migration and behaviour of lowland river fishes at a fish pass. Fisheries Research. 1999;44(2):183–191.

[30] Kesner BR, Karam AP, Pacey CA, Marsh PC. Demographics and post-stocking survival of repatriated razorback sucker in Lake Mohave. 2010 annual report. Boulder City, NV: Bureau of Reclamation Agreement No. R09AP30002; 2010. 32 p.

[31] Snyder DE. Electrofishing and its harmful effects on fish. Information and technology report USGS/BRD/ITR-2003-002. Denver, CO: U.S. Government Printing Office; 2003. 149 p.

[32] U.S. Fish and Wildlife Service. Management plan for the big river fishes of the Colorado River basin: amendment and supplement of the bonytail, humpback chub, Colorado pikeminnow, and razorback sucker recovery plans. Albuquerque, NM: Department of the Interior; 2005.

[33] Braun DP, McClure CJW. Razorback sucker (*Xyrauchen texanus*) (RASU) basic conceptual ecological model for the lower Colorado River. Boulder City, NV: Lower Colorado River Multi-Species Conservation Program; 2013. 184 p.

[34] Lower Colorado River Multi-Species Conservation Program. Final Science Strategy. Boulder City, NV: Lower Colorado River Multi-Species Conservation Program; 2007. 71 p.

Chapter 8

Chemometric Analysis of Wetlands Remnants of the Former Texcoco Lake: A Multivariate Approach

Jacinto Elías Sedeño-Díaz,
Alexis Joseph Rodríguez-Romero,
Erick Mendoza-Martínez and Eugenia López-López

Additional information is available at the end of the chapter

http://dx.doi.org/10.5772/63436

Abstract

The former Texcoco Lake (TL) belongs to a closed basin (Basin of México), with a semi-arid climate and soil types Solonchaks and Vertisol, which confer saline and alkaline properties to the waterbodies. Historically, this lake has been facing extraction of salt for agricultural and livestock use. As a result of population growth, the lake has lost more than 95% of its surface; however, currently TL is formed by permanent and temporary shallow wetlands interconnected by the rainy season which are used as buffer zones to avoid possible floods to the City of Mexico and as receptors of wastewater discharges. Due to the above, the former TL has a very diverse mosaic of conditions in the remaining wetlands, therefore their quality assessment comprises a complex task of interpretation and analysis. To perform this, it is important to have a series of indicators and analysis tools that take into account the state of contamination of the waterbodies in different periods of study. Additionally, is imperative to have essential elements of interpretation which allow to combine all the features of analysis in a comprehensive scrutiny of the health status of these water bodies. The purpose of this research is to analyze the quality of water and sediments of the remaining wetlands located in the basin of the former TL through various physicochemical parameters and heavy metals, and using multivariate methods and water and sediment quality indices.

Keywords: saline lake, water wells quality, chemometric, sediment quality, wetlands

1. Introduction

Due to constant and rapid human population growth, water bodies around the world have been threatened [1]; even large lakes are considered especially vulnerable because they are the final receivers of a number of pollutants, in most cases due to poor connectivity with other waterbodies, these pollutants are deposited indefinitely altering the quality of water and sediment [2]. Among the main activities affecting lentic systems worldwide are overexploitation of water and other resources associated with water bodies, contamination by direct or indirect sources, agriculture, and industry [3]. It has been estimated that coverage occupying lakes and ponds around the world ranges from 1.3 to 1.8% and although it is numerically dominated by small systems, the largest contribution in terms of the total area is given by the great lakes [4–7].

Another important aspect regarding the current status of epicontinental waterbodies is the distribution of freshwater lakes in relation to salt water. According to [8], about half of the epicontinental water of the world is saline (in area and volume), while the other half is freshwater; an epicontinental waterbody is considered saline if it exceeds the salt concentration of 3 g/L without marine environment contact [9]. Regarding its location, epicontinental saline lakes are commonly found in endorheic basins, arid or semiarid climates, and geographically within a latitudinal range of 20–40° in both hemispheres [6, 10].

Mexico has 0.1% of epicontinental water of the world and more than half a million cubic meters is represented by lakes [8]. Saline lakes in Mexico are located mainly in the north and center of the country associated with arid and low rainfall [6, 11]. In this sense, one of the most important saline lakes is the former Lake Texcoco, which has been a great biological and cultural referring in Mexico; Texcoco Lake and lakes of Zumpango, Xaltocan, Xochimilco and Chalco formed a large lake complex where the ancient Aztec civilization was established [12].

The former Texcoco Lake belongs to a closed basin (Basin of México), with a semiarid climate and summer rains, with maximum temperature ranges from 30 to 32°C in April and June, while the minimum from 2 to 5°C in October to March. The soil in the lake are Solonchaks and Vertisol, which give saline and alkaline properties to the waterbodies [10], favoring the concentration of ions of manganese, ammonium, nitrites, nitrates, sulfates, and metals [12]. According to [13], this lake is athalassohaline and alkaliphilic due the high concentration of ions such as magnesium, sulfate, calcium, sodium, and bicarbonate derived from rocks and geological weathering. Historically, this lake has been used for the extraction of salt from the ancient Aztec civilization; later, with the arrival of the Spaniards it has been continuously drained by agriculture, livestock, and expansion of the City of Mexico, losing more than 95% of its surface [6, 12, 14].

Nowadays, this lake has become a fragmented aquatic ecosystem, formed by the remaining temporary or semi-permanent wetlands interconnected during the rainy season whose function has been to avoid possible floods to the City of Mexico, as well as to act as a wastewater receptor. Because of this, in the former Lake Texcoco, there is a very diverse mosaic of environmental conditions in the remaining wetlands; thus, the assessment of water and sediment quality comprises a complex task of interpretation and analysis. To perform this task, it is important to use a series of indicators and analysis tools that account the assessment of the state of contamination of the waterbodies in different periods of study, using elements of multivariate analysis, which allow the joint interpretation of the parameters analyzed for a comprehensive scrutiny of the health status of these water bodies. In this sense, the aim of this research is to analyze the current condition of water quality of the remaining wetlands located in the basin of the former Texcoco Lake through numerous physical and chemical parameters and heavy metals, using multivariate methods and indices of water quality.

2. Chemometric analysis and multivariate approach

One of the great challenges in the aquatic ecosystems assessment is the opportunity of integrating natural phenomena and pollution that occur in rivers and lakes [15]. The elements of variation in a waterbody can be discrete or specific depending on the nature of the study site [16]; having a data set that reflects clearly the main trends and sources of pollution or variation of physical, chemical, and biological properties in these environments is crucial.

In this sense, chemometric analysis can be defined as a discipline derived from chemistry that uses mathematics, statistics, and logic to design or select the optimal experimental procedures that provide the most relevant chemical information data analysis for obtaining knowledge from natural systems [17]. This tool in conjunction with multivariate analysis is able to reduce the multidimensionality of the parameters in natural environments to get graphical models or indices that provide insight into the current state of water bodies [18]. In recent years, the application of different statistics multivariate techniques, such as cluster analysis (AC), principal component analysis (PCA), factor analysis (FA), and discriminant analysis (AD) have contributed to identifying possible sources that influence waterbodies, and have offered a valuable tool in the reliable management of water resources [19, 20]. These techniques have been successfully used to classify water quality data and detect similarities between samples and/or parameters in many research studies [21, 22].

3. Study area

The area of the former Texcoco Lake is located between 19° 20′N, 98° 05′W and 19° 35′N, 99° 05′W (**Figure 1**). The lake is an area where Vertisol, Solonchak, and the saline soil type predominates [23], which characterizes this as one of the few saline lakes of Mexico. It is located at an average altitude of 2200 m, with an average annual temperature of 16°C and rainfall of 705 mm [24].

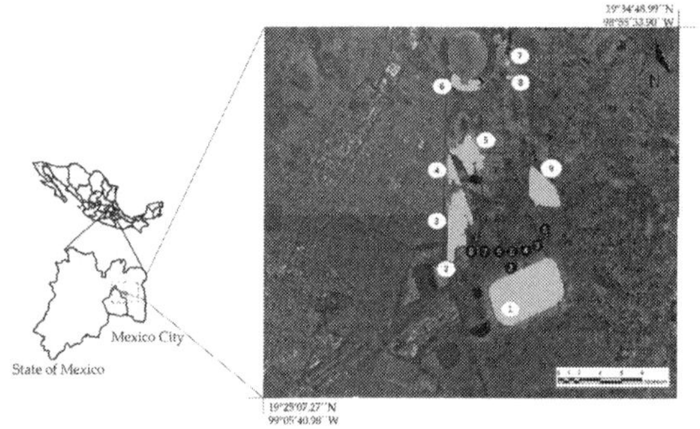

Figure 1. Location of waterbodies and wells studied in the former Texcoco Lake. The white circles represent waterbodies: 1. Lago Nabor Carrillo; 2. Laguna Cola de Pato; 3. Laguna de Evaporación; 4. Laguna Colorada; 5. Laguna 14; 6. Depósito de Evaporación; 7. Laguna 16; 8. Casa del Ermitaño and 9. Xalapango. The black circles represent wells: 1. Vivero; 2. Colector múltiple del Lago Nabor Carrillo; 3. Well No. 15; 4. Well No. 16; 5. Well No. 17; 6. Well No. 18; 7. Well No. 19 and 8. Well No. 21.

The waterbodies in the area of the former Lake Texcoco include a system of interconnected wetlands in an interrupted and ephemeral mode with a heterogeneity of uses including historical uses. The Lago Nabor Carrillo is an artificial waterbody that receives input from treated water from local plants and from zones of the east of Mexico City. On the other hand, the evaporation tank (Depósito de evaporación) is an artificial water body constructed for extraction of caustic soda, currently out of operation. In contrast, within these water bodies are the Laguna 16, the House of Hermit (Casa del Ermitaño), and Xalapango that are natural water bodies receiving contributions from rivers coming from the northeast and southeast regions adjacent to the area of Texcoco Lake.

In relation to wells, the depth of static water level fluctuates from 40 to 130 m [25]. The shallowest values are recorded in the southwestern portion of the aquifer near Lago Nabor Carrillo, which gradually increase by topographic effect from north and east, where the foothills of the mountains that delimit the aquifer are located.

4. Methods

4.1. Sampling methods and analytical procedures

Water samples were collected from 19 study sites from nine waterbodies, as well as a set of eight wells (**Figure 1**) in the area of the former Lake Texcoco in June, September, and November 2015. Water samples were taken at surface using plastic containers of 2.5 L and free from metal and detergents following the methods of field sampling for waterbodies receiving wastewater and sampling of water for human consumption proposed by [26].

All water samples were kept at 4°C in dark conditions for transportation for further laboratory analysis.

Water quality analyses included 25 physicochemical parameters (**Table 1**), of which only temperature, conductivity, pH, and dissolved oxygen were measured *in situ*. Likewise, ten metals were quantified in water, sediments, and wells. Analytical procedures were performed in accordance with the guidelines of Mexican standards [27, 28].

Physicochemical parameters		**Sample**
Temperature (°C) (T°C)	Hardness (mg/L) (Dur)	Waterbody and water well
pH	Phosphates (mg/L) (PO_4)	
Dissolved oxygen (mg/L) (OD)	Nitrates (mg/L) (NO_3)	
Conductivity (μS/cm) (Cond)	Nitrites (mg/L) (NO_2)	
Total solids (mg/L) (ST)	Total nitrogen (mg/L) (TN)	
Total suspended solids (mg/L) (SST)	Ammonia nitrogen (mg/L) (NH_3)	
Total dissolved solids (mg/L) (SDT)	Organic nitrogen (mg/L) (ON)	
Settleable solids (mg/L) (SD)	Oil and grease (mg/L) (O and G)	
Turbidity (NTU) (Turb)	Sulfates (mg/L) (SO_4)	
Biochemical oxygen demand (mg/L) (BOD)	Methylene blue active substances (mg/L) (MBAS)	
Chemical oxygen demand (mg/L) (COD)	Total coliforms (NMP/100 ml) (Tot Col.)	
Cyanides (mg/L) (CN)	Fecal coliforms (NMP/100 ml) (Fec Col.)	
Chlorides (mg/L) (Cl)		
Metals		
Aluminum (mg/L) (Al)	Cadmium (mg/L) (Cd)	Waterbody, sediment and water well
Lead (mg/L) (Pb)	Chrome (mg/L) (Cr)	
Iron (mg/L) (Fe)	Nickel (mg/L) (Ni)	
Arsenic (mg/L) (As)	Copper (mg/L) (Cu)	
Barium (mg/L) (Ba)	Mercury(mg/L) (Hg)	

Table 1. Physicochemical and metals analyzed in the water and sediment samples.

4.2. Analysis and data interpretation

Trophic state index (TSI) [29] was calculated from the results of the total phosphorus concentration and using the following mathematical expression:

$$TSI(PT) = 10[6 - \frac{ln\frac{48}{TP}}{ln2}]$$

where *TSI (PT)* is the trophic state index for total phosphorus and *TP* is the concentration of total phosphorus (mg m^{-3}). The index expresses the trophic state based on the following scale: <30 as oligotrophic; 40–50 mesotrophic; eutrophic 50–70; and 70–100 or greater than 100 as hypereutrophic [29].

On the other hand, the Water Quality Index proposed by the National Sanitation Foundation (WQI-NSF) [30] generates results from 0 to 100, with zero considered the worst quality and 100 as excellent quality. The WQI-NSF has the following expression:

$$WQI = \sum_{i=1}^{9} WiTi(pi) = \sum_{i=1}^{9} wiqi$$

where p is measured for the ith parameter; Ti is the transformed value from the curve pi quality of each parameter, such that $Ti(pi) = qi$ and wi are the relative weight for the ith parameter.

$$\sum_{i=1}^{9} wi = 1$$

The WQI-NSF represents water quality in general giving different grades depending on the value obtained, but without incorporating qualifications for specific water uses, such as drinking water, recreational use, suitable for agriculture, etc.

Finally, the proposed Pollution Index (Cd) by [31] was calculated, which determines the relative contamination of different metals separately and presents the sum generated by each of these components. This contamination index is calculated according to the following equation:

$$C_d = \sum_{i=1}^{n} C_{fi} \quad (1)$$

$$C_{fi} = \left(\frac{C_{Ai}}{C_{Ni}}\right) - 1 \quad (2)$$

where $C_f i$ is the contamination factor for the metal, $C_A i$ the analytical result of the metal, $C_N i$ the maximum permissible limit of the metal. According to [32]

Pollution levels are referred to *Cd*:

$Cd < 1$ Low contamination

$1 < Cd < 3$ Medium contamination

$Cd > 3$ High contamination

Additionally, in the case of sediments, [33] proposed the metal pollution index (MPI), which is a very simple index, calculated as a geometric mean concentration of metals by the expression:

$$MPI = (M_1 \cdot M_2 \cdot M_3 \cdot \ldots \cdot M_n)^{\frac{1}{n}}$$

This index has the peculiarity that does not use reference levels, safety data sheets, reference sites, or baselines. There is no classification of polluted or unpolluted, but allows comparisons with other aggregations (sites) and highlighting those sites with high concentrations of metals.

The variation of each parameter over the periods and study sites was performed by a factor analysis (FA) to identify which parameters are statistically significant for further analysis. From this information, a cluster analysis (AC) was assessed using the Euclidean Distances Index and Ward ligation method.

The set of indicators of water quality analysis was performed by Pearson correlation in a PCA from a data matrix consisting of the study sites as rows and environmental parameters as columns. The array data was standardized by $X = \log(x + 1)$ to make evident the chemometric trends of water bodies from the former Texcoco Lake. All statistical analyses were performed using the software XLSTAT-2015.

5. Results

5.1. Physico-chemical characterization of the waterbodies from the former Lake Texcoco

The water bodies of the former Texcoco Lake have a high conductivity with values ranging between 932 and 12.266 mS/cm, mainly due to the high concentration of chlorides and metals such as Fe and Mg. On the other hand, the concentration of total suspended solids, dissolved solids, and settleable solids is very high, with values of up to 88.39 mg/L in the study sites: S2 (Laguna Cola de Pato) and S5 (Laguna 14); these conditions are due to the shallowness of the waterbodies and because the allochthonous materials entering the waterbodies are easily deposited. Furthermore, the high content of dissolved material in the water increases the concentration of BOD_5 and COD, reaching values of up to 195.76 and 7,858 mg/L respectively, in S2 and S5 sites.

5.2. Factor analysis

The FA for the set of physicochemical parameters showed that the first factor F1 (51.15% of the explained variance) includes the parameters related to the input of organic compounds, such as dissolved oxygen, BOD_5, COD, total nitrogen, ammonia nitrogen, and organic nitrogen, as well as elements that provide ions and suspended materials such as conductivity, total solids, fluorides, phosphates, sulfates, and MBSA. In the second factor F2 (20.62% of the explained variance), the factors related to the obstruction in light penetration were pondered such as suspended solids, turbidity, oil, and grease (**Table 2**).

	F1	**F2**
Temperature (°C)	**0.694**	−0.310
pH	**0.840**	−0.162
Dissolved oxygen (mg/L)	**0.614**	−0.443
Conductivity (μS/cm)	**0.880**	−0.435
Total solids (mg/L)	**0.892**	−0.246
Suspended solids T (mg/L)	0.304	**0.813**
Dissolved solids T (mg/L)	**0.845**	−0.410
Settleable solids (mL/L)	−0.474	0.357
Turbidity (NTU)	0.386	**0.827**
BOD soluble (mg/L)	**0.799**	0.213
COD soluble (mg/L)	**0.923**	−0.026
Chlorides (mg/L)	**0.896**	−0.391
Hardness (mg/L)	0.288	−0.187
Fluorides (mg/L)	**0.890**	−0.214
Phosphates (mg/L)	**0.807**	−0.214
Nitrates (mg/L)	0.153	0.388
Nitrites (mg/L)	0.203	0.513
Total nitrogen (mg/L)	**0.913**	0.342
Ammonia nitrogen (mg/L)	**0.715**	0.677
Organic nitrogen (mg/L)	**0.927**	0.220
Oil and grease (mg/L)	0.436	**0.751**
Sulfates (mg/L)	**0.664**	−0.645
MBAS (mg/L)	**0.769**	0.141
Total coliforms (NMP/100 mL)	−0.084	**0.866**
Fecal coliforms (NMP/100 mL)	**0.683**	0.181
Solonchak	**0.920**	0.315
Vertisol	**−0.920**	−0.315
Variability (%)	**51.153**	20.620
% accumulated	**51.153**	**71.773**

F1 and F2 belong to factorial loads of FA. Bold number indicates significant values.

Table 2. Factor analysis (FA) for the periods and study sites and the environmental parameters.

5.3. Cluster analysis

The nine water bodies were separated into two groups and four independent units (**Figure 2**). The first group corresponds to the study sites S3 (Laguna Evaporation), S1 (Lago Nabor Carrillo, the largest waterbody and receiving wastewater), and S4 (Laguna Colorada); these sites are in the south and southeast of the former Texcoco Lake. The second group corresponds to the study sites S2 (Laguna Cola de Pato) and S5 (Laguna 14), which are located in the northwest region of the former Texcoco Lake. These waterbodies have the highest conductivity values, suspended materials (ST, SST, SDT and SD), BOD_5, and COD. Finally, the four independent units correspond to S6 (Depósito de evaporación), S7 (Laguna 16), S8 (Casa del Ermitaño), and S9 (Xalapango). These sites are in the north and northeastern portion of the lake.

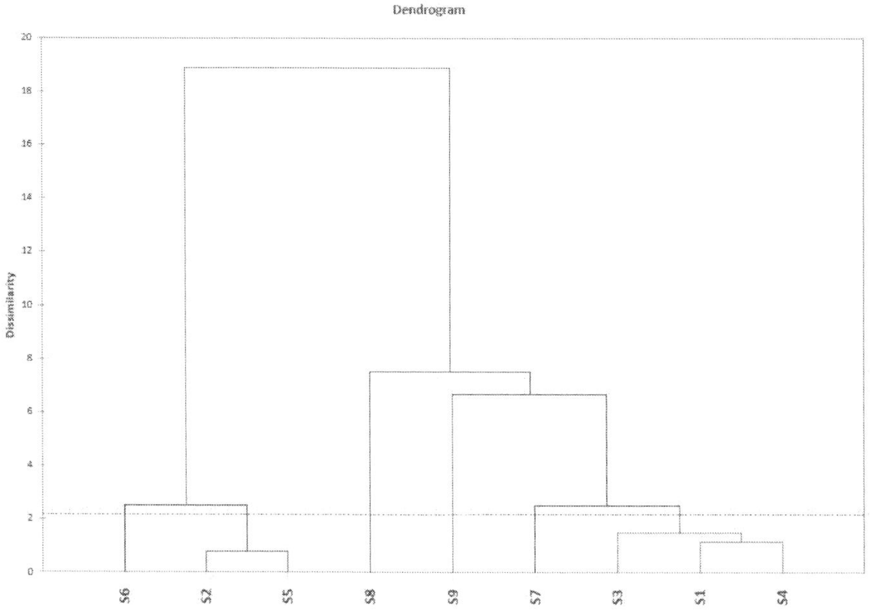

Figure 2. Dendrogram of study sites with the Euclidean distances dissimilarity index, based on the parameters of greater significance in the factor analysis.

5.4. Water quality index

The WQI-NSF scores ranged from 18.51 in S3 in September to 44.19 in S7 during June. However, all scores in the WQI-NSF qualify from waters with poor quality (41 to 50) to very poor quality (0–40) (**Figure 3a**).

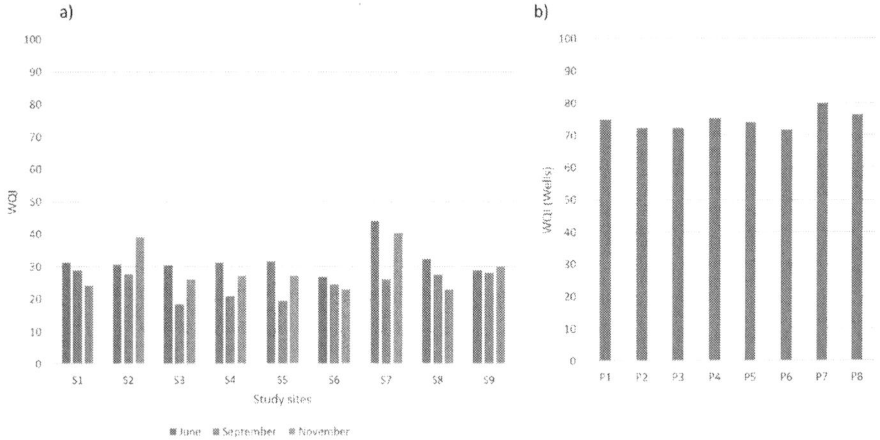

Figure 3. Water Quality index (WQI-NSF) values: (a) WQI-NSF for the waterbodies during the three periods of study; (b) WQI-NSF for the wells in the former Texcoco Lake.

Mean values of WQI for all study sites and periods of study are 32.39, 24.71, and 28.95 in June, September, and November, respectively. However, the dispersion of the data in September is wider, indicating low and high values of WQI-NSF in some water bodies.

According to the records of the National Meteorological System [34], the rainy season of this region ends in September (96.6 mm/month), which could contribute to the depletion of WQI-NSF by the increased input of allochthonous materials and increased wastewater flow from the rivers of the eastern region of the former Texcoco Lake. Similarly, the runoff in a closed basin directly increases settleable and suspended solids in the waterbodies, causing further deterioration of water quality.

According to [14, 35], the rivers from the eastern portion of the former Texcoco Lake provide to these water bodies high values of organic matter and fecal contamination, reaching 49.75 million fecal coliforms/100 mL^{-1}. High values of coliform not only exceed the maximum permissible guidelines of the Mexican standards (1000–2000 NMP 100 mL^{-1}), but also considerably deplete the results of water quality index.

5.5. Trophic state index

The TSI values fluctuated between 106.19 in site S8 to 180.43 in site S4 during June. All sites in the three periods of study showed hypereutrophic condition (70–100+). However, it is remarkable that waterbodies S2 and S5 showed the highest values during November, where the concentration of PO_4 reached 156.8 mg/L and 107.98 mg/L, respectively, with the highest values of this nutrient throughout the study sites and seasons (**Figure 4**).

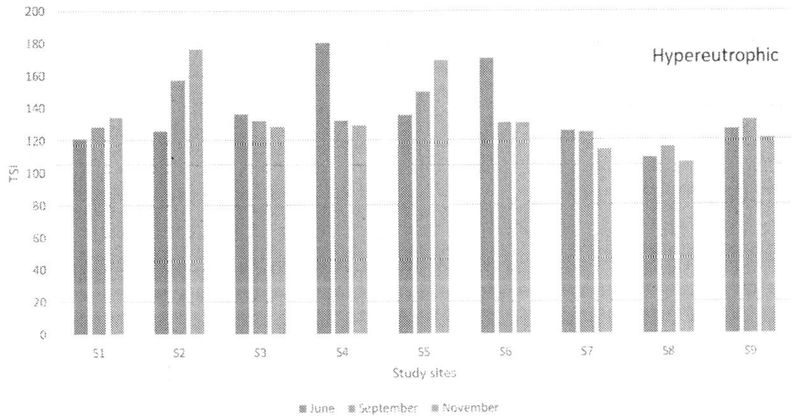

Figure 4. Values of TSI for water bodies during the three periods of study.

5.6. Pollution Index

Results of Pollution Index show that study sites S1, S3, S7, and S8 showed values from low pollution ($Cd < 1$) to medium pollution ($1 < Cd < 3$), which is consistent because the site S1 receives treated wastewater, while the sites S7 and S8 are small waterbodies isolated without input from wastewater (**Figure 5**).

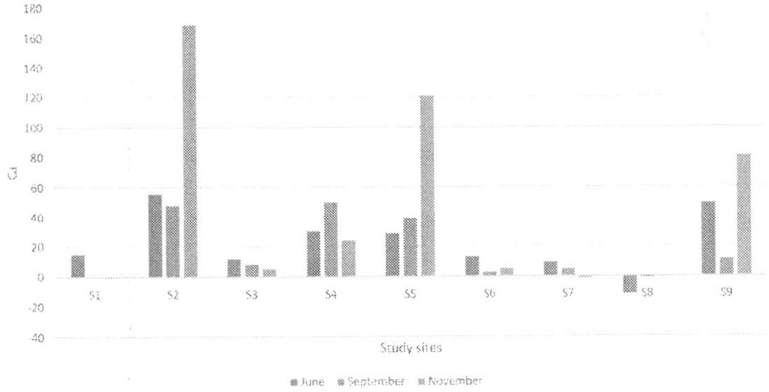

Figure 5. Values of the metal pollution index (Cd) for the nine bodies of water during three study periods.

In contrast, other sites qualify with high pollution, being the site S2 (168.75), S5 (121.01) and S9 (80.43) in November those sites that reached the highest values of metal contamination. In the case of S9, despite being a natural water body, it receives inputs from rivers located in the eastern portion of the former Texcoco Lake and according to [14, 35], they transport in their current a considerable amount of contaminants and heavy metals such as Pb, Zn, Ni, Cu, and Cd, although the concentrations do not exceed the maximum permissible guidelines.

The high metal concentrations, particularly for waterbodies of the central part of the lake, are because many of these waterbodies are small and very shallow, with a high variation in volume. On the other hand, continuous evaporation processes concentrates salts in the water, since they have also the highest concentrations of conductivity (>20,000 mS/cm) chlorides (>120,000 mg/L) and total dissolved solids (>300,000.00 mg/L). Finally, being small and shallow waterbodies, they are influenced by Solonchak soil that predominates in the area. In these soils, aluminum salts (alumina) are predominant, which may be contributing to the increase in the index scores as the aluminum concentration exceeds more than forty times the reference value.

5.7. Metal pollution index

The MPI provides a geometric mean of the metal concentration for each waterbody, thus this index is an indicator of the level of contamination by these elements. There are metals whose presence in nature or polluting sources is very low, while other metals may be in abundance (either in nature or polluting sources). The assessment of geometric mean values contributes to the elimination of those extreme values that affect the correct interpretation of the data.

The result set of geometric mean values was considered as the overall average for all the water bodies studied (**Figure 6a** and **b**, horizontal orange line). The global geometric mean value for all the study sites indicates that waterbodies S2 and S5 are those with the highest values of MPI for the three periods of study, with values of 0.54, 0.57, and 0.77 and 0.50, 0.41 and 0.71 for the months of June, September, and November, respectively. This coincides with the Pollution Index, which indicates that these waterbodies are the most contaminated by metals.

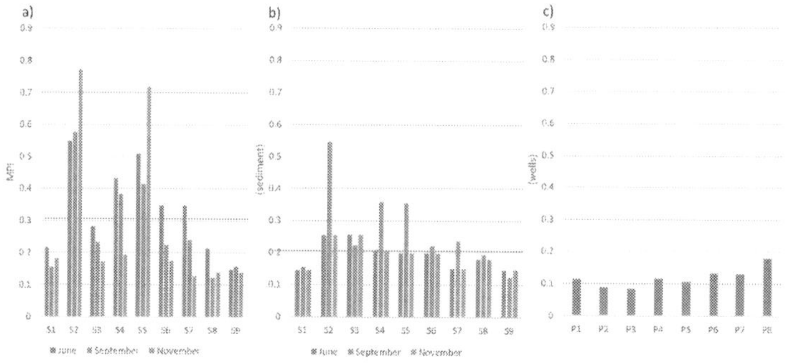

Figure 6. Values of the metal pollution index (MPI) for the nine water bodies during the three periods of study (a); for sediments of nine water bodies (b) and for the eight wells (c).

According to this index, the sites S7, S8, and S1 were sites with lower concentration of metals during the three periods of study, which in relation to the previous analysis allows to group these waterbodies with the best quality indicators among all water bodies studied.

5.8. Metal contamination in sediments

In the case of sediments, the MPI showed a maximum value of 0.54 for S2 in September, while the minimum value was 0.12 for S9 during September. The overall mean was 0.21 (geometric mean of all water bodies and periods of study) (**Figure 6b**). It is a remarkable increase of metals in sediments during September, which may be due to the effects of the rains that carry metals to waterbodies and these can be settled down with the material being transported, or can be chelated and removed from the water column to the sediments. This analysis shows that there is a higher concentration of metals in water reaching values up to 0.77-fold, related to the concentration in sediments.

5.9. Water quality of wells

The scores for the wells qualify as good water quality (**Figure 3b**). According to the National Water Commission [36], the static level of depth of the wells for the area of Texcoco fluctuates between 40 and 130 m, the wells being closest to Lago Nabor Carrillo with depths of 50 m [37]. The geological profile of the Texcoco area consists of clays with a high concentration of salts and silts up to 80 m deep [35], giving rise to very poor permeability of surface water into the aquifer.

Furthermore, the MPI scores for wells reflect the scarce metal concentration (**Figure 6c**). However, the high concentration of Mn (0.17–1.86 mg/L) is a distinctive feature of the geology of the area and in this sense, the presence of this metal is removed by a local treatment plant that recovers water from this series of wells for public supply. This suggests that the quality of water in the aquifer in this area can be considered good, although the geological nature and soil type provide metals and a large amount of salts.

6. Integration of the indices

In the present study, we used four indicators of pollution of water bodies:

- Water quality index (WQI-NSF).
- Trophic state index (TSI).
- Cd pollution index (indicator of contamination by heavy metals based on international security standards for different uses of water).
- Metal pollution index (MPI).

The WQI-NSF is an indicator of water quality by different environmental parameters measured in the water and allows decisions about the use of the resource; in this case, all water bodies were classified into two categories: (a) poor quality and (b) very poor quality.

The TSI is an indicator of nutrient enrichment, which indicates that due to the high concentration of nutrients these waterbodies are susceptible to algal blooms, both by microalgae or cyanobacteria, which can become potentially toxic.

Cd pollution index is based on comparative environmental standards. Its purpose is to ensure environmental safety standards regarding the use of water. In this case we used international criteria for aquatic life protection, protection of human health and public supply, of which the strictest criteria were selected.

The MPI is indicator of the level of metals in water and sediments, offering an overall average metal content in each compartment. Based on the above, a principal component analysis (PCA) (**Figure 7**) was performed, and to identify trends, three of the five indices were used: TSI, WQI-NSF, and MPI; thus the analysis considers nutrient enrichment, the state of water quality and metal contamination.

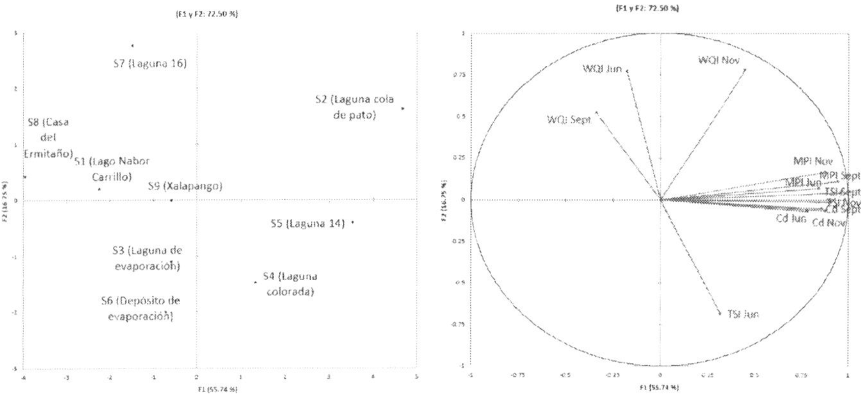

Figure 7. Biplot of the Principal Component Analysis for the water bodies based on TSI, WQI-NSF, Cd pollution index, and MPI.

The use of all these indices allows the identification of different kinds of deterioration of the waterbodies.

The PCA showed a 72.50% explained variance, with 55.74% in the first component and 16.75% in the second. On the upper and lower right quadrants the vectors of MPI, TSI, and Cd were located and the three sampling seasons were associated with these vectors. Furthermore, the waterbodies S2, S4, and S5 were also associated with these vectors. These waterbodies, as was already indicated, have the highest concentrations of metals. Similarly, S2 and S5 reached the highest values of the TSI for the three periods of study.

On the upper left quadrant, the vector of WQI-NSF showed that sites S1, S7, S8, and S9 in June and September reached the highest values of WQI; in addition, these sites exhibit the lowest concentration of metals and lowest values of the TSI. In this sense, these water bodies are characterized by better water quality among all the study sites.

7. Conclusions

Water bodies within the ancient Texococo Lake, assessed with different tools, show a high degree of contamination.

The WQI-NSF, an indicator of physical and chemical quality, notes that water bodies showed poor to very poor water quality.

All water bodies have a high degree of nutrient enrichment and are in hypereutrophic conditions.

The concentrations of heavy metals through various indices (Cd and MPI) showed that the 20 waterbodies are highly contaminated by metals, with a higher concentration in both water and sediments.

Water quality of wells is good despite the high concentration of Mn by natural geological effects. Also by the geological features of the study area, the pollution of the surface water does not compromise the quality of the aquifer, mainly due to low permeability of the lithologic material in this area.

The integration of the WQI-NSF, TSI, Cd, and MPI indices and use of multivariate analysis identified those bodies of water in which metal contamination is more relevant than the deterioration of water quality by other physico-chemical parameters, as well as those water-bodies that have better conditions, despite their poor water quality.

Acknowledgements

The authors thank the Instituto Politécnico Nacional for the facilities offered for this research.

Author details

Jacinto Elías Sedeño-Díaz[1*], Alexis Joseph Rodríguez-Romero[2], Erick Mendoza-Martínez[2] and Eugenia López-López[2]

*Address all correspondence to: jsedeno@ipn.mx

1 Polytechnical Coordination for Sustainability, National Polytechnic Institute, Coordinación Politécnica para la Sustentabilidad, Instituto Politécnico Nacional, Av. IPN esq. Wilfrido Massieu s/n, Ciudad de México, México

2 Laboratory of Aquatic Ecosystems Health Assessment, National School of Biological Sciences, National Polytechnic Institute, Laboratorio de Evaluación de la Salud de Ecosistemas Acuáticos, Departamento de Zoología, Escuela Nacional de Ciencias Biológicas, Instituto Politécnico Nacional, Ciudad de México, México

References

[1] Boyd CE. Water Quality: An Introduction. 2nd ed. USA: Springer; 2014. 355 p. DOI: 10.1007/978-3-319-17446-4

[2] William M, Lewis Jr. Basis for the protection and management of tropical lakes. Lakes and Reservoirs: Research and Management. 200;5(1):35–48. DOI: 10.1046/j.1440-1770.2000.00091.x

[3] Fleeger WJ, Carman KR, Nisbet RM. Indirect effects of contaminants in aquatic ecosystems. The Science of the Total Environment. 2003;317(1):207–233. DOI: 10.1016/S0048-9697(03)00141-4

[4] Wetzel RG. Wetlands as metabolic gates. Great Lakes Research. 1992;18(4):529–532. DOI: 10.1016/S0380-1330(92)71320-3

[5] Meybeck M. Physics and Chemistry of Lakes. 1st ed. Berlin: Springer; 1995. 1–35 p. DOI: 10.1007/978-3-642-85132-2_1

[6] Alcocer J, Hammer UT. Saline lake ecosystems of Mexico. Aquatic Ecosystem Health and Management. 1998;1(3–4):291–315. DOI: 10.1080/14634989808656925

[7] Dowing JA, Prairie YT, Cole JJ, Duarte CM, Tranvik LJ, Striegl RG, et al. The global abundance and size distribution of lakes, ponds, and impoundments. Association for the Sciences of Limnology and Oceanography. 2006;51(5):2388–2397. DOI: 10.4319/lo.2006.51.5.2388

[8] Alcocer J, Bernal-Brooks FW. Limnology in Mexico. Hydrobiologia. 2010;644(1):15–68. DOI: 10.1007/s10750-010-0211-1

[9] William WD. Environmental threats to salt lakes and the likely status of inland saline ecosystems in 2025. Environmental Conservation. 2002;1(2):154–167. DOI: 10.1017/S0343560602000103

[10] Fernández-Buces N, Siebe C, Cram S, Palacio JL. Mapping soil salinity using a combined spectral response index for bare soil and vegetation: a case study in the former lake Texcoco, Mexico. Arid Environments. 2006;65(4):644–667. DOI: 10.1016/j.jaridenv.2005.08.005

[11] Spring OU. Aquatic systems and water security in the Metropolitan Valley of Mexico City. Current Opinion in Environmental Sustainability. 2011;3(6):497–505. DOI: 10.1016/j.cosust.2011.11.002

[12] González-Morán T, Rodríguez R, Cortes SA. The basin of Mexico and its metropolitan area: water abstraction and related environmental problems. South American Earth Sciences. 1999;12(6):607–613. DOI: 10.1016/S0895-9811(99)00043-7

[13] Litchfield CD. Saline lakes. In: Reitner J, Thiel V. Encyclopedia of Geobiology. United States of America: Springer Science+Business; 2011. pp. 765–769. DOI: 10.1007/978-1-4020-9212-1

[14] Guzmán-Quintero O, Palacios-Vélez OL, Carrilo-Gonzáles R, Chávez-Morales J, Nikolskii-Gavrilov I. The contamination of superficial water in the basin of river Texcoco, México. Agrociencia. 2007;41(4):385–393.

[15] Kebede YK, Kebedee T. Application of principal component analysis in surface water quality monitoring. In: Sanguansat P, editor. Prinicpal Component Analysis—Engineering Applications. Thailand: InTech; 2012. pp. 83–100. DOI: 10.5772/38049

[16] Sheela AM, Letha J, Joseph S., Chacko M, Kumar S, Thomas J. Water quality assessment of a tropical coastal lake system using multivariate cluster, principal component and factor analysis. Lakes and Reservoirs: Research and Management. 2012;17(2):143–159. DOI: 10.1111/j.1440-1770.2012.00506.x

[17] Massart DL, Vabdeginste BGM, Buydens S, De Jong PJ, Smeyers-Verbeke J. Introduction. In: Massart DL, Vabdeginste BGM, Buydens S, De Jong PJ, Smeyers-Verbeke J, editors. Data Handling in Science and Technology Handbook of Chemometrics and Qualimetrics Part: A. 38th ed. Amsterdam: Elsevier; 1998. pp. 1–20. DOI: 10.1016/S0922-3487(97)80031-7

[18] Zhao Y, Xia XH, Yang ZF, Wang F. Assessment of water quality in Baiyangdian Lake using multivariate statistical techniques. In: Yang Z, Chen B, editors. 18th Biennial ISEM Conference on Ecological Modelling For Global Change and Coupled Human and Natural System 2012; 20–23 September 2012; Beijing, China. China: Procedia Environmental Sciences; 2012. pp. 1213–1226. DOI: 10.1016/j.proenv.2012.01.115

[19] Reghunath R, Seredhara-Murthy TR, Raghavan BR. The utility of multivariate statistical techniques in hydrogeochemical studies: an example from Karnataka, India. Water Research. 2002;36(10):2437–2442. DOI: 10.1016/S0043-1354(01)00490-0

[20] Simeonova P, Simeonov V, Andreev G. Water quality study of the Struma river Basin, Bulgaria (1989–1998). Open Chemistry. 2003;1(2):121–136. DOI: 10.2478/BF02479264

[21] Kotti ME, Vlessidis AG, Thanasoulias NC, Evmirids NP. Assessment of river water quality in Northwestern Greece. Water Resources Management. 2005;19(1):77–94. DOI: 10.1007/s11269-005-0294-z

[22] Nakano T, Tayasu I, Yamada Y, Hosono T, Igeta T, Hyodo A, et al. Effect of agriculture on water quality of Lake Biwa tributaries, Japan. Science of the Total Environment. 2008;389(1):132–148. DOI: 10.1016/j.scitotenv.2007.08.042

[23] Moreno-Sánchez E. environmental and sociopolitical characteristics of Texcoco municipality. Mexico State. Quivera., 2007;08.042.

[24] Ruiz-Romero E, Alacántara-Hernández R, Cruz-Mondragon C, Marsch R, Luna-Guido ML, Dendooven L. Denitrification in extreme alkaline saline soils of the former lake Texcoco. Plant and Soil. 2009;319(1–2):247–257. DOI: 10.1007/s11104-008-9867-y

[25] National Commission of Water (CONAGUA). Update of the annual medium availability, water in the Texcoco (1507), Mexico State [Internet]. 20/04/2015.

[26] Federation WE, American Public Health Association. (2005). Standard Methods for the Examination of Water and Wastewater. Washington, DC, USA: American Public Health Association (APHA).

[27] Secretary of Environment and Natural Resources (SEMARNAT). NOM-001-SEMARNAT-1996 [Internet]. 30-10-1996 [Updated: 30-10-1996]. Available from: http://www.ordenjuridico.gob.mx/Documentos/Federal/wo69205.pdf [Accessed: 14-03-2016]

[28] Secretary of Environment and Natural Resources (SEMARNAT). NOM-002-SEMARNAT-1996. [Internet]. 03-06-1998 [Updated: 03-06-1998]. Available from: http://www.aguascalientes.gob.mx/PROESPA/pdf/NOM-SEMARNAT-002%20DESCARGA%20DE%20AGUAS%20RESIDUALES%20AL%20ALCANTARILLADO.pdf [Accessed: 14-03-2016]

[29] Carlson RE. A trophic state index for lakes. Limnology and Oceanography. 1977;22(2): 361–369. DOI: 10.4319/lo.1977.22.2.0361

[30] Brown RM, McClelland, NI, Deininger RA, Tozer RG. A water quality index: do we dare?. Water & Sewage Works. 1970;117:339–343.

[31] Backam B, Bodis D, Lahermo P, Rapant S, Tarvainen T. Application of a groundwater contamination index in Finland and Slovakia. Environmental Geology. 1998;36(1):55–64. DOI: 10.1007/s002540050320

[32] World Health Organization (WHO). Reference Guides Water Quality. [Internet]. 2016. Available from: http://www.who.int/water_sanitation_health/dwq/guidelines/es/ [Accessed: 14-03-2016]

[33] Usero J, González-Regalado E, García I. Trace metals in the bivalve mollusc *Chamelea gallina* from the Atlantic coast of southern Spain. Marine Pollution Bulletin. 1996;32(3): 305–310. DOI: 10.1016/0025-326X(95)00209-6

[34] National Meteorology Service (CONAGUA). Mexico State [Internet]. 2016. Available from: http://smn.cna.gob.mx/index.php?option=com_content&view=article&id=189:estado-de-mexico&catid=14:normales-por-estacion [Accessed: 14-03-2016]

[35] Rivera-Vázquez R, Palacios-Vélez OL, Chavéz-Morales J, Belmont MA, Nikolskii-Gavrilov I, De la Isla De Bauer MaL, et al. Contamination or Coliforms and Helmints in the rivers of Texcoco Chapingo and San Bernadito, tributaries of the eastern part of the Valley's Mexico Basin. Internacional of Water Contamination. 2007;23(2):69–77.

[36] Water National Comision (CONAGUA). Project of artificial water recharge of the Experimental Aquifer Module of Tertiary Treatment in the Federal zone of Texococo lake [Internet]. 06-2011 [Updated: 05-2012]. Available from: http://cuencavalledemexico.com/wp-content/uploads/2012/05/GEO-78-Recarga-Artificial-Texcoco-IG-UNAM1.pdf [Accessed: 14-03-2016]

[37] Escobar-Villagrán BS, Palacios-Vélez OL. Analysis of overexploitation of Texcoco aquifer, Mexico. Tecnology and Sciences of water. 2012;3(2):67–84.

Chapter 9

Fuzzy Logic as a Tool for the Assessment of Water Quality for Reservoirs: A Regional Perspective (Lerma River Basin, Mexico)

Jacinto Elías Sedeño-Díaz and Eugenia López-López

Additional information is available at the end of the chapter

http://dx.doi.org/10.5772/64265

Abstract

The aim of this study is to propose a water quality index for reservoirs in a basin using fuzzy logic. Most of the water quality indices are designed for use in rivers and streams and based on expert opinion; however, when the water is dammed, the quality usually Is modified. Mexico is a country with many contrasts in quantity and quality of water. Management of water resources in the Lerma River is achieved with a system of artificial reservoirs where water is stored in order to meet human needs, such as public supply, industry, agriculture and recreation, among others. Monitoring of 11 reservoirs in the Río Lerma basin was performed to characterize the water quality. Using the water quality data, those indicators that do not represent redundancy were selected based on the concentration gradient that occurred in the different reservoirs. Thus, the proposed index uses eight indicators of water quality. The fuzzy inference system is composed by 633 rules with a score from 0 to 100 and seven verbal categories. The index was validated by comparison with other water quality index, and their use across the basin was tested by applying it in five additional water bodies.

Keywords: fuzzy inference system, physicochemical approach, reservoirs, water quality

1. Introduction

Water quality assessment traditionally has been measured by physical and chemical parameters and through the comparison of results of monitoring programs with the existing local guidelines. Only in some of these cases, the use of this methodology allows for a proper identification of contamination causes and may be essential for checking legal compliance; however, it does not readily give a global behaviour of the water quality in a basin [1].

Water quality indices have been used to translate large data sets on water quality into a single value representing a certain level of water quality [2]. The common denominator for all water quality indices is its ability to combine data from monitoring programs by means of a simple quality vector [3].

The most common water quality constituents used in water quality indices include dissolved oxygen (DO), pH, temperature, faecal coliforms, turbidity, biochemical oxygen demand, chlorides, colour, nitrates, total phosphorus, and total solids [3, 4]. These parameters traditionally have been selected and qualified through the combined judgment of a panel of water quality experts within a region or country utilizing a set of questionnaires based on the Delphi system [5, 6].

Water quality indices are mainly used by water resource managers to communicate whether water is acceptable for its intended uses [3, 6–9], as well as to compare and identify trends between different watersheds or water bodies, and facilitate comparison among different sampling sites and/or events [1, 2, 10, 11].

In addition to water uses (public water supply, agricultural irrigation, industrial, navigation, recreation, etc.), which are qualified by water quality indices, the services provided by water bodies, such as maintenance of groundwater level, balance of atmospheric gasses, climate regulation, and reduction of soil erosion, are extremely important to human welfare; however, their importance has decreased and is not considered in the assessment tools. There is evidence that many human dominated ecosystems have become highly stressed and dysfunctional [12].

The evolution of water quality indices has been satisfactory in terms of water resources management because they were all developed for a specific set of goals such as rating the water use [13], communication tool, and decision-making managers [6], and they are based on criteria or standards (environmental benchmarks) that reflect the impairment of quality caused by the presence of pollutant/parameter considered in the water quality index (WQI).

Horton [14] is considered the pioneer in the design of water quality indices, and he proposed that various water quality characteristics could be integrated into an overall index. This first WQI was defined as a rating reflecting the composite influence on overall quality of a number of individual quality characteristics. After Horton, numerous indices have been developed that include different water quality characteristics, calculation methods, and different purposes. Dinius [6] proposed a geometric WQI qualifying until six water uses.

Numerous studies have used water quality indices to determine water quality in rivers [2, 3, 9, 15–20]. Recently, fuzzy theory has been used to design water quality indices and also to assess the water quality in rivers [21–25].

Few studies, much less at a regional level, have been conducted to assess the water quality in lakes or reservoirs in terms of a water quality index. In some cases, studies are available on trophic state development [26–29]. López [28] carried out a regional study in reservoirs of the Río Lerma basin, pointing out the existence of a trophic gradient from a regional perspective.

Sedeño-Díaz and López-López [30] applied a geometric WQI to 10 reservoirs of the Río Lerma basin, and Shuhaimi-Othman et al. [31] carried out a study of the water quality changes of Chini Lake. Fuzzy synthetic evaluation [32] and fuzzy theory [33, 34] have been utilized for diagnostic of trophic state in reservoirs. None of these WQI has been developed from a gradient of environmental conditions inside a hydrologic basin which allows to compare the different water bodies inside the same region.

Human population growth and changes in adjacent land use have increased the pollutant and nutrient inputs in the reservoirs, altering water quality and accelerating the eutrophication of reservoirs, lakes and watercourses. The Río Lerma basin with an extension of only 3% of the Mexican territory is the most important water system in the central plateau of Mexico. Likewise, urban areas, agricultural lands, and industrial centres located along its course are set to become one of the most densely populated and polluted regions in the country. This basin has experienced negative impacts due to human activities; it currently faces an imbalance between water demand and availability, primarily due to its natural water scarcity as well as uneven water quality distribution. The rapid urban and industrial growth among other economic and social factors has made this worse. Water needs have grown, water users are fiercely competing with each other and conflicts are emerging as a result. Hence, water quality has also deteriorated as urban and industrial effluents are often discharged without treatment. The Río Lerma basin is also considered as a centre for fish fauna endemism [2] and therefore is mandatory to take conservation measures.

To overcome the water availability problem, numerous reservoirs have been built to satisfy the needs of the population. Cotler-Ávalos [35] indicates that at present there are 552 reservoirs in the basin. Therefore, it is important to have simple and easy-to-use tools to assess the water quality of the reservoirs and facilitate interpretation and decision making, since they are the main source of water to meet the needs of the population in that region. In this study, a water quality index (WQI) based on fuzzy logic was designed to assess and to compare the environmental condition of several reservoirs of the Río Lerma basin, using a selection of eight water quality characteristics.

2. Background about fuzzy logic and fuzzy inference systems

The aim of this section is not to expose the full fuzzy logic theory (FL); however, it is important to give a brief introduction. FL was introduced in 1965 by [36], and it is a mathematical tool for dealing with uncertainty as it is able to measure linguistic concepts or subjective words that are fundamentally imprecise, ambiguous or fuzzy [21, 37].

We can ask what does a water quality index do with the FL. For several years, FL has been applied to design environmental indices because it solves complex situations such as ambiguity, subjective judgments, and interpretation of a complex set of multidimensional data [22, 37]. The results of a WQI are most often associated with different linguistic water quality categories (e.g. excellent, good, regular, or bad water quality). These linguistic variables use unclear boundaries, that is, these terms include a high degree of uncertainty. In addition, considerable vagueness is involved in the allocation of a water quality score for multiple uses inclusive of a specific use [11]. FL can be considered as a language that allows us to translate the uncertainty of natural language into mathematical expressions [11]. Thus, FL has been considered as a useful tool for modelling water quality as it is an alternate approach to problems where the goals and boundaries are diffuse or imprecise [24, 38].

2.1. Fuzzy inference systems (FISs)

FISs are based on the fuzzy set theory, which maps input values to output values [23, 38]. The input is called *antecedent*, while the output is known as *consequent*. Maps are outlined in the membership functions.

A membership function is a curve whose shape is defined by convenience [11, 38], and that defines how each point in the antecedent is mapped to a membership value in a range of 0–1 [21]. In FIS, different shapes of membership functions can be used, such as Gaussian, bell, trapezoidal and triangular, among others. Trapezoidal and triangular membership functions have the advantage of being asymmetric [39], but the gradient of values of membership develops over the same slope value.

In concordance with [11, 38, 40, 41], a FIS consist of three main steps:

- Fuzzification, is the process which changes a crisp input data to a fuzzy number expressed in a membership function, that is, the transformation of a numerical value of any water quality variable into a membership grade to a fuzzy set.
- Evaluation of fuzzy decision through the system of linguistic *If-Then* rules which include the fuzzy operators to integrate the combined antecedents to the consequent.
- Defuzzification, is the process to obtain a representative value of a fuzzy set, that is, the final crisp value that integrates all attributes of the multiple antecedents. There are different methods of defuzzification, the most common are centroid, mean of maxima, and bisector; however, it is very important to select an appropriate defuzzification method.

These fundamental three steps are imperative to obtain a successful FIS.

In concordance with [38], there are at least six reasons to use models based on fuzzy rules and fuzzy sets: (a) they can be used to describe a large variety of nonlinear relations, (b) they tend to be simple, since they are based on a set of local simple models, (c) they can be interpreted verbally and this makes them analogous to artificial intelligence models, (d) they use information that other methods cannot include, (e) the fuzzy approach has a big advantage over other indices as they have the ability to expand and combine quantitative and qualitative data that express the water quality status, and finally, (f) FL can deal with and process missing data without compromising the final result.

http://dx.doi.org/10.5772/64265

3. Methods

3.1. Study area

We studied 11 reservoirs with different water use (power generation, agriculture irrigation, drinking trough, recreational, and public supply) and different location within the Río Lerma basin (upper, middle, and lower reaches), all of them considered as hydrological priority systems by Mexican Environmental Authorities (**Figure 1**).

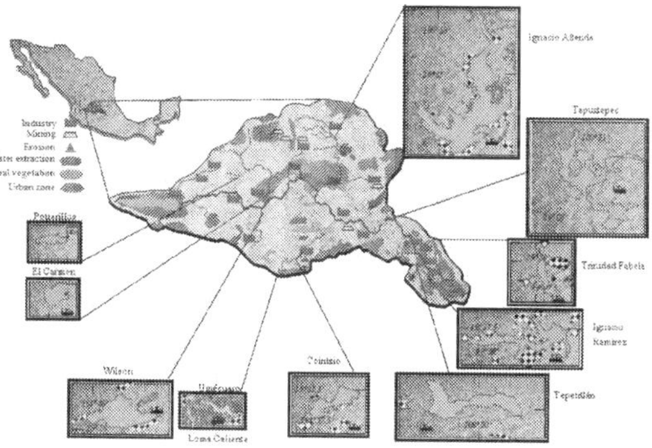

Figure 1. Río Lerma basin and location of reservoirs studied.

The upper Río Lerma includes the following reservoirs: Ignacio Ramírez (IR), Tepetitlán (TP), Trinidad Fabela (TF), and Tepuxtepec (TX). In the middle portion of Río Lerma are Ignacio Allende (IA), Potrerillos (PO), Umecuaro (U), Loma Caliente (LC), Carmen (CA), and Cointzio (CO), and finally, the reservoir studied in the lower Río Lerma was Wilson (W).

Some reservoirs are located in the headstreams of some Río Lerma tributaries (LC, U, TP, TF, PO), while others are close to urban or industrial centres (IA, CO, TX) or adjacent to agriculture and livestock areas (CA, W, IR). Nonetheless, they all sustain human influence of one kind or another.

3.2. Water quality variables

Water quality for each reservoir was characterized by means of 19 parameters, four times in an annual cycle to determine the spatial and temporal variation in one year: dissolved oxygen (mg/L), water temperature (T, °C), Secchi disk transparency (SDT, m), chlorophyll *a* (Chl *a*, μg/L), turbidity (Turb, NTF), and specific conductance (Cond, μS/cm) were measured *in situ* with a Hydrolab DataSonde Surveyor 4, while biochemical oxygen demand (BOD, mg/L), total nitrogen (TN, mg/L), nitrates (NO_3, mg/L), nitrites (NO_2, mg/L), ammonia (NH_4, mg/L), total phosphorus (TP, mg/L), ortho-phosphates (O-PO_4, mg/L), total suspended solids (TSS, mg/L), and colour (C, Pt-Co units) were determined through Hach techniques with a Hach spectrophotometer DR2500. Alkalinity (Alk, mg/L), hardness (H, mg/L), and chlorides (Cl, mg/L) were determined by titration, and finally, total dissolved solids (TDS, mg/L) were measured with a TDS meter Hach model 44600.

A multivariate discriminants analysis (DA) was applied to water quality data to find trends and reservoirs sharing similar characteristics. Maximum, minimum, and median of all water quality parameters along 11 reservoirs were taken into account to select those parameters that should be incorporated in the FIS. Likewise, multiple regressions were carried out to find relationships among different parameters and to eliminate those with redundancy or without significance.

The range of values for each selected parameter was considered by taking into account the absolute lowest and the absolute highest values in all the reservoirs and is expressed along the x-axis in the membership function curve.

The membership functions were assigned using the Gaussian curve because is the shape that better reflects the semantic meaning of each parameter considering that increased or decreased water quality is not lineal.

Linguistic classification for each water quality parameter in the antecedent was considered only with three categories: Excellent, Medium, and Bad. In the case of water temperature, only two categories were used: Excellent and Not Excellent, since both low and high temperatures alter the physiology of organisms inhabiting the water of the reservoirs. The linguistic variables in the consequent output were considered as Unacceptable, Very Polluted, Contaminated, Regular Quality, Slightly Polluted, Good Quality, and Excellent.

Different defuzzification methods were tested (Bisector; Centroid; Large of Maximum, LOM; and Middle of Maximum, MOM) to select the best method.

In addition, the water quality index proposed by [6] was computed for all the reservoirs to obtain a reference value of water quality (benchmark).

3.3. Mathematical analyses

Statistical analyses were performed using the StatistiXL version 1.8. The Fuzzy Water Quality Index for the Río Lerma reservoirs was carried out using the Fuzzy Logic Design Toolbox of MATLAB V. R2013a.

4. Results

4.1. Water quality index as reference status

As a first approach, WQI proposed by [6] (WQI_{Dinius}) was computed for all the reservoirs to have a reference status of water quality. Mean values for each reservoir are show in **Table 1** and **Figure 14**. The maximum value of WQI was in LC, a reservoir located in a headstream of a tributary of Río Lerma. The lowest score of WQI was detected in TX, which is located downstream of an urban and industrial zone. These results are the benchmark to compare the new water quality index.

http://dx.doi.org/10.5772/64265

Reservoir	LC	U	TF	TP	PO	W	CO	IR	CA	IA	TX
WQI	79.9	74.83	71.54	70.94	69.84	67.78	67.21	66.42	62.82	62.79	62.49

Table 1. Mean WQI_{Dinius} scores for all the reservoirs.

4.2. Selection of environmental variables

In order to select the environmental variables (water quality parameters) to be used in the setting of the new Fuzzy Water Quality Index for reservoirs of Río Lerma basin ($FWQI_{Lerma}$), a DA was performed with the purpose to detect groups of reservoirs sharing water quality characteristics. In this sense, DA scatter plot showed a significant formation of four groups (Wilk's Lambda = 15E-8, $p < 0.001$, **Figure 2**). Using box and whisker plots, we detect the environmental variables that typify the groups of reservoirs as follows:

Group I:Reservoirs (U, LC, and Pot) with SDT > 0.5 m (**Figure 3a**), and the lowest concentration of TSS (**Figure 3b**), turbidity (**Figure 3c**), colour (**Figure 3d**), conductivity (**Figure 3g**), and TDS (**Figure 3h**).

Group II:Reservoirs (Tepe and TF) with the lowest concentration of nutrients (nitrates and ortho-phosphates, **Figures 3g** and **3h**, respectively).

Group III:Reservoirs (Co and W) with the highest concentration of nitrates (**Figure 3g**), turbidity (**Figure 3c**), and colour (due the presence of clay, **Figure 3d**), and the lowest values of hardness (**Figure 3i**).

Group IV:It includes the reservoirs IR, Car, IA, and Tepu, which shows the highest concentration of hardness (**Figure 3i**), TSS (**Figure 3b**), ortho-phosphates (**Figure 3f**), biochemical oxygen demand (**Figure 3j**), and the lowest values of SDT (**Figure 3a**).

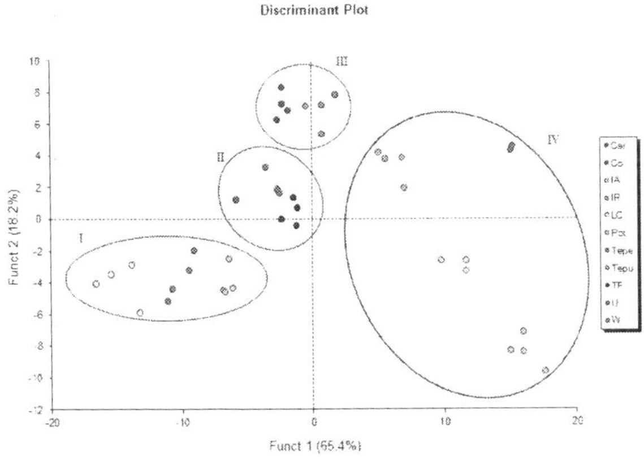

Figure 2. Scatter plot of the discriminant analysis of reservoirs based on their water quality attributes.

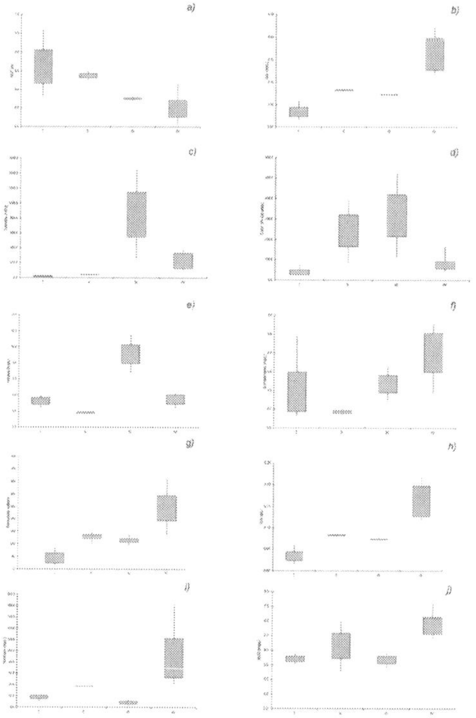

Figure 3. Box and whisker plots of physicochemical variables that characterize each group of reservoirs according to DA.

Based on these groups, five variables were selected: (a) Secchi disk transparency, (b) conductivity, (c) nitrates, (d) ortho-phosphates, and (e) colour, with the following justification:

Secchi disk transparency: This is an important physical parameter in lentic systems because it has a close relationship with turbidity (physical and biological), the total suspended solids and colloidal particles. In several cases, there is a direct relationship with chlorophyll *a* content and therefore, in such cases, can be an indicator of biological productivity for lentic systems.

In the Río Lerma basin, a relationship between SDT, turbidity, Chl *a*, and TSS data was determined for all the water bodies studied. The following expression summarizes the relationship among these parameters:

$$SDT = 1.83 - 0.00598 ln(Turbidity) - 0.27 ln(Chl\,a) - 0.101 ln(TSS) \qquad R^2 = 0.816$$

Thus, we can consider only SDT measure as a representative parameter of Chl *a*, turbidity, and TSS — the last two are parameters that characterize the reservoirs according to DA.

http://dx.doi.org/10.5772/64265

The membership function for SDT was obtained considering the minimum (0.07 m) and maximum value (1.5 m) of transparency detected in all the reservoirs (**Figure 4**).

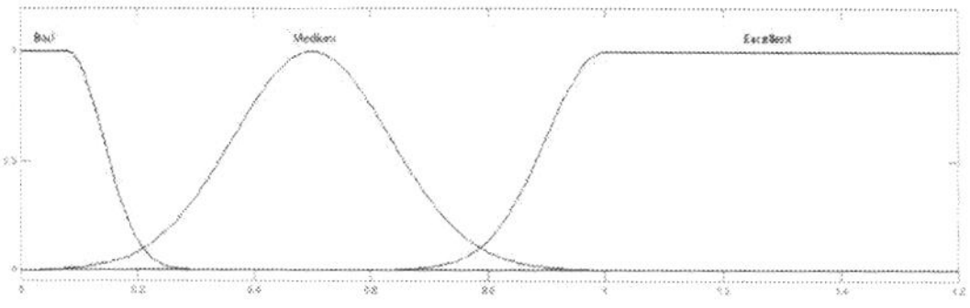

Figure 4. Membership function for SDT.

Conductivity: The specific conductance or conductivity represents the salinity of water. It is a measure of the ability of water to conduct electrical current; likewise, conductance qualitatively reflects the status of inorganic pollution and is a measure of total dissolved solids and ionised species in the water [16, 19]. An empirical relationship between total dissolved solids and conductivity can be derived for any stream. High levels of dissolved and suspended solids in the water systems increase the biological and chemical oxygen demand, which deplete the dissolved oxygen levels in the aquatic systems. The levels of TDS in a broad sense reflect the pollutant burden of the aquatic system [16], and include the carbonates and sulphates that are considered in hardness measurements. Therefore, conductivity is an important parameter to be considered in the $FWQI_{Lerma}$. The membership function was considered taking into account a range of values of conductivity from 0 to 700 μs/cm (**Figure 5**).

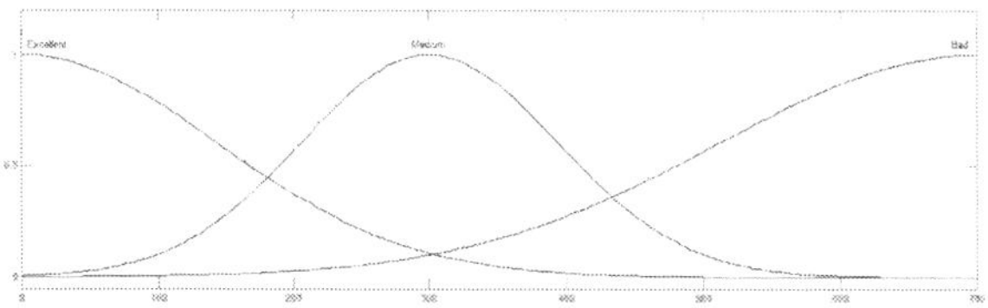

Figure 5. Membership function for conductivity.

Colour: Water colour is indicative of substances in solution or in colloidal suspension, but also is the result of interplay of light on suspended particulate materials together with such factors as bottom or sky reflection. Dissolved substances and particulate organic matter contribute to the colour and turbidity of natural waters. It is also indicative of algae blooms [7]. For the Río Lerma basin reservoirs, in addition to SDT, colour was considered as an important factor due the nature of the substrate, because colour varies according to the type of clays found in different regions of the basin independently of primary production due to algae. Reservoirs Co and W were those with the highest values of colour due to clays. The membership function of colour was determined considering a range of values from 0 to 700 Pt-Co units (**Figure 6**).

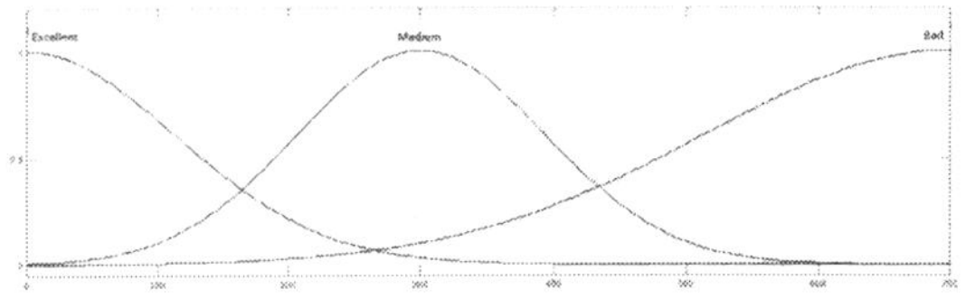

Figure 6. Membership function for colour.

Ortho-phosphates: They are the bioavailable chemical species of phosphorus for the aquatic organisms, which is the main reason for its consideration in this index. Still more than total phosphorus, this one can be in nonbioavailable dissolved inorganic forms for the organisms or in particulated form (like part of the aquatic organisms). Furthermore, ortho-phosphates are an indicator of the trophic state, as well as of municipal effluents and the agricultural runoff. The membership function was performed considering a range from 0 to 12 mg/L of ortho-phosphates (**Figure 7**).

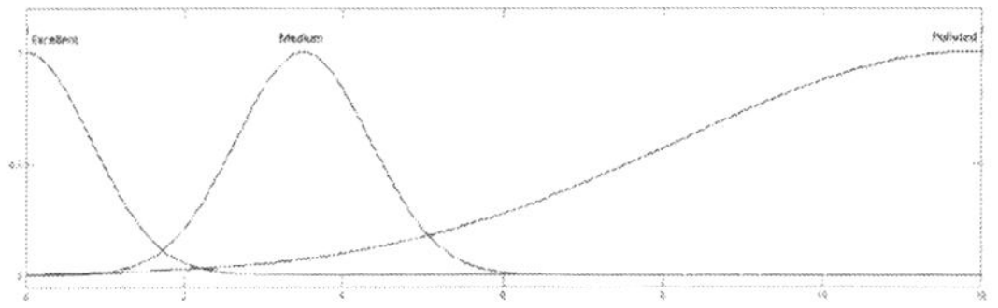

Figure 7. Membership function for ortho-phosphates.

Nitrates: The nitrates are a chemical species of the nitrogen bioavailable to be used by the aquatic biota, mainly by the primary producers. Nitrates are a source of nitrogen present in water column that permits the aquatic biota to cover their nutritive needs of nitrogen. Likewise, it is an indicator related to the trophic state in lentic systems. The membership function was performed considering a range of values from 0 to 40 mg/L, taking into account all the reservoirs (**Figure 8**).

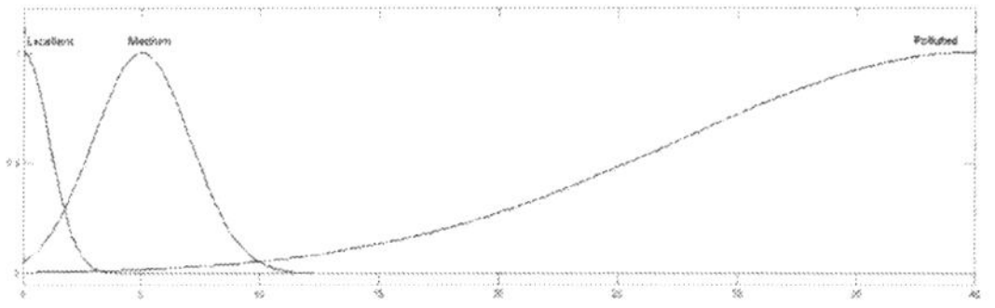

Figure 8. Membership function for nitrates.

Three additional variables were included in the $FWQI_{Lerma}$: dissolved oxygen, water temperature, and ammonia, taking into account the following:

Dissolved oxygen is one of the critical parameters for aquatic life support and the most frequently measured parameter in monitoring studies. This parameter represents the amount of oxygen that is available to aquatic organisms for metabolism/respiration and assimilation of food [42]. DO is an indicator of photosynthetic activity and the deoxygenation and reaeration factors such as water currents, temperature, wave action, and other disturbances at the reservoir surface results in a greater passage of the oxygen into solution. Membership function was based on the percentage of saturation of DO, taking into account that temperature and altitude are the principal factors that affect the concentration of DO, and that 100% represents the better condition of DO (**Figure 9**).

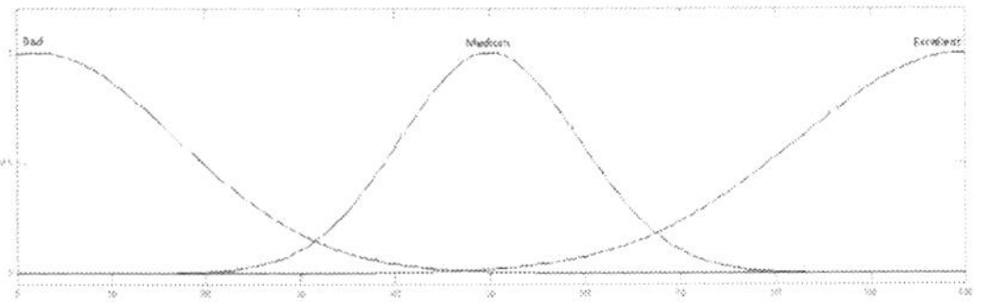

Figure 9. Membership function for DO.

Water temperature is an important parameter in water quality because it has a great relationship with the physiology of the aquatic organisms; in lentic systems, the temperature shows a gradient in the first metres of deep, and in certain reservoirs, a thermocline may occur. To determine a value of temperature, which represents an ideal value for the fish fauna and other aquatic organisms living in the Río Lerma basin, the mean value of temperature from 1975 through 1999 in the 17 monitoring stations (in all of the three portions of this basin) was considered as a satisfactory value; the data were taken from [2]. The 100% of membership (excellent) was adjusted on the mean value of water temperature in that period. Temperature values above or below the mean value were considered "Not Excellent" in a gradient of decrease or increase (**Figure 10**).

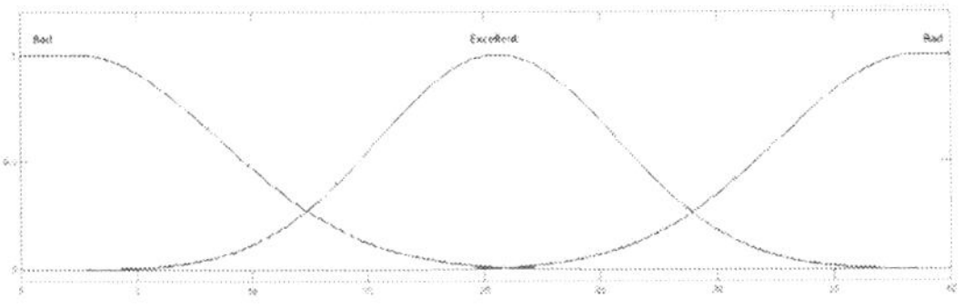

Figure 10. Membership function for water temperature.

Ammonia is considered because it is a chemical that participates in the cycle of nutrients (N) and is an indicator of organic pollution that is faster and easier to determine than the total and faecal coliforms. It is an indicator of faecal pollution and municipal wastewaters. Ammonia is excreted by animals and is produced during decomposition of plants and animals. Ammonia is an component in many fertilizers and is also present in sewage, storm water runoff, certain industrial wastewaters, and runoff from animal feedlots. Furthermore, ammonia can be toxic depending on the temperature and pH. In this sense, ammonia can be an excellent water quality indicator of organic pollution, mainly of domestic wastewater. The membership function was considered with a range from 0 to 1.6 mg/L, which were the values observed through all the reservoirs (**Figure 11**).

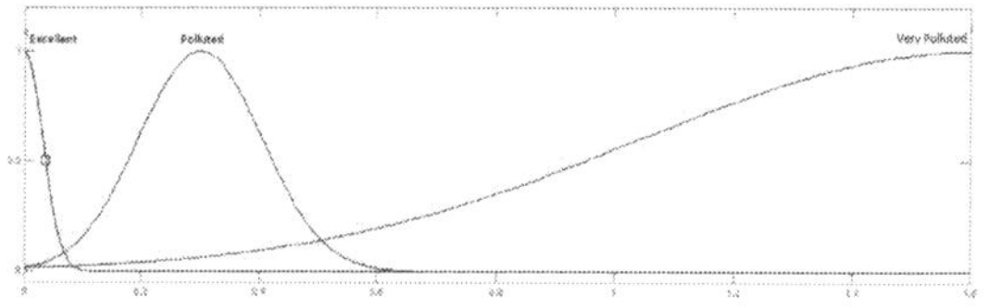

Figure 11. Membership function for ammonia.

4.3. Inference rules (*If-Then*)

Since eight water quality variables were used to perform the $FWQI_{Lerma}$, it was necessary to conform a system of *If-Then* rules. The effect of the different water quality variables cannot be isolated because all of them occur simultaneously in the water. Then, a single rule must incorporate all the variables, and so all the rules. Thus, the FIS for $FWQI_{Lerma}$ was composed by 633 *If-Then* rules considering the eight water quality variables as the antecedent and seven linguistic categories in the consequent output.

Examples of inference rules for each category of the consequent output are:

- If DO is bad, and NO_3 is excellent, and DS is medium, and Specific Cond is excellent, and O-PO4 is excellent, and Colour is excellent, and Ammonia is excellent, and Temp is not Excellent, then WQI is *Unacceptable*.
- *If* DO is medium, and NO_3 is medium, and DS is bad, and Specific Cond is bad, and O-PO_4 is bad, and Colour is bad, and Ammonia is bad, and Temp is excellent, *then* WQI is *Very Polluted*.
- *If* DO is excellent and NO_3 is bad, and DS is bad, and Specific Cond is medium, and O-PO_4 is medium, and Colour is bad, and Ammonia is bad, and Temp is Excellent, *then* WQI is *Contaminated*.
- *If* DO is excellent and NO_3 is excellent, and DS is bad, and Specific Cond is medium, and O-PO_4 is bad, and Colour is bad, and Ammonia is medium, and Temp is Not excellent, *then* WQI is *Regular Quality*.
- *If* DO is excellent and NO_3 is excellent, and DS is excellent, and Specific Cond is medium, and O-PO_4 is medium, and Colour is medium, and Ammonia is medium, and Temp is excellent, *then* WQI is *Slightly contaminated*.

- *If* DO is excellent and NO_3 is excellent, and DS is excellent, and Specific Cond is excellent, and O-PO_4 is medium, and Colour is medium, and Ammonia is Good Quality, and Temp is excellent, *then* WQI is *Good Quality*.
- *If* DO is medium and NO_3 is excellent, and DS is excellent, and Specific Cond is excellent, and O-PO_4 is excellent, and Colour is excellent, and Ammonia is excellent, and Temp is excellent, *then* WQI is *Excellent*.

The consequent output is a crisp value as a result of defuzzification process, which is associated with the linguistic category in the consequent. Thus, the Input-Output map of the FIS for $FWQI_{Lerma}$ is depicted in **Figure 12**.

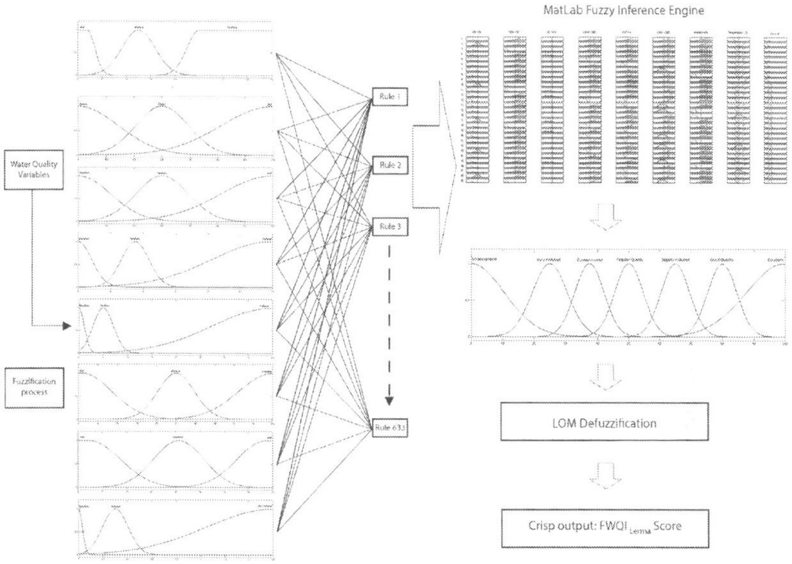

Figure 12. Map for fuzzy water quality index inference system.

4.4. Selection of defuzzification method

As indicated above, different methods of defuzzification were tested, for which ANOVA was performed between the $WQI_{Diniuis}$ results and the scores obtained for $FWQI_{Lerma}$ with the application of the following defuzzification methods: Bisector, Centroid, LOM, and MOM. In this case, the best method was selected based on the minimum statistical difference between WQI proposed by [6] (benchmark) and defuzzification results. **Figure 13** shows the box and whisker plot of the comparison between WQI_{Dinius} score and the scores of the different methods of defuzzification; LOM is the one with the smallest difference with WQI_{Dinius}; in fact, there is a total overlap. In this sense, LOM was selected as the method of defuzzification for $FWQI_{Lerma}$.

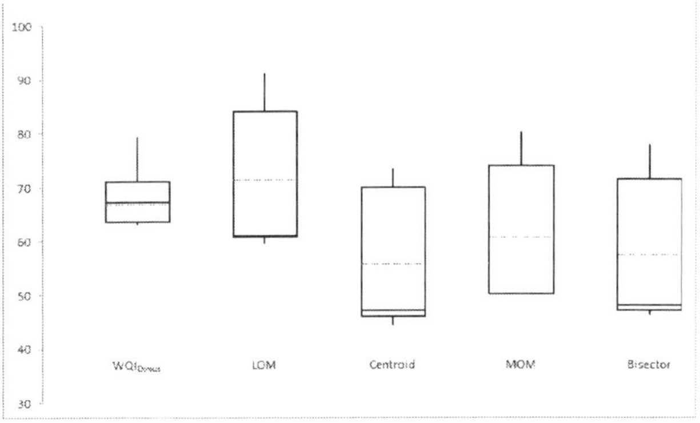

Figure 13. Box and whisker plot representing the results of ANOVA of the application of different defuzzification methods compared with the benchmark WQI_{Dinius}.

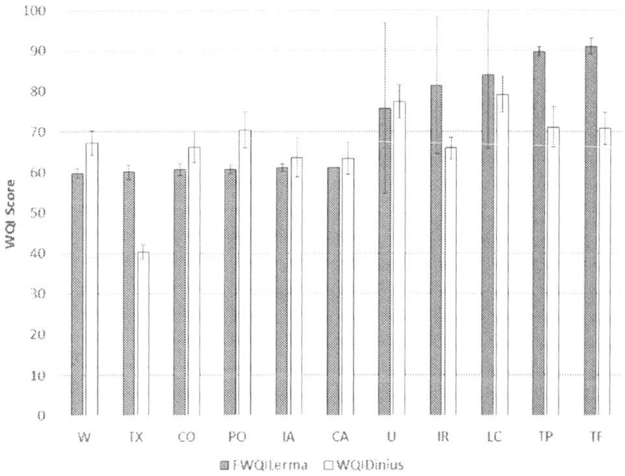

Figure 14. WQI scores (±SD) of Río Lerma reservoirs, considering results of WQI_{Dinius} and $FWQI_{Lerma}$.

Once the defuzzification method was selected, it is possible to compute the $FWQI_{Lerma}$ for all the reservoirs. **Figure 14** shows the WQI scores for WQI_{Dinius} and $FWQI_{Lerma}$. The best scores were obtained from LC, U, TP, and TF reservoirs, which are in headstreams.

4.5. Model validation

In order to identify the effectiveness of the $FWQI_{Lerma}$, a validation process was carried out using five new water bodies located in different sites into the Río Lerma basin. Thus, two water bodies with an excellent water quality were selected: Nieves and Zacapu Lake; on the other hand, additional water bodies were selected with a regular water quality: Melchor Ocampo Reservoir, Pool Lake, and Solis Reservoir.

When the $FWQI_{Lerma}$ was applied to these water bodies, the first two showed a score of 86, and the other three obtained the score of 64, 64, and 62, respectively (**Figure 15**), showing that $FWQI_{Lerma}$ effectively reflects the water quality status in other water bodies of the same basin.

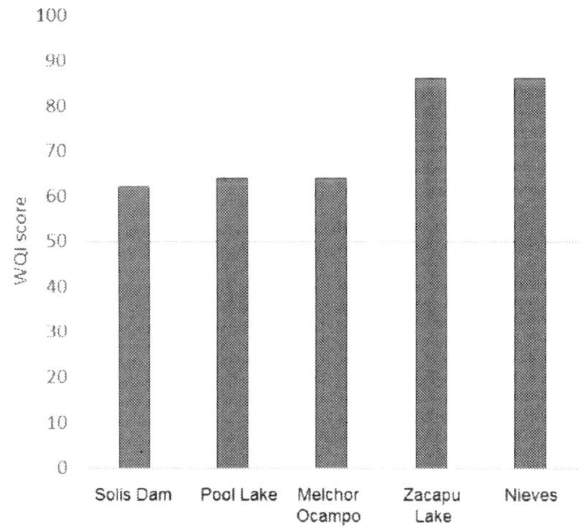

Figure 15. WQI scores for water bodies of validation.

5. Discussion

The WQI have been an excellent tool to assess water quality using physicochemical approach. Historically, water quality indices have been applied by environmental agencies to take decisions about water management and conservation and to advise the water quality status to the public. Both regional (ecoregions) and basin approaches have proven to be the most successful tools for the assessment of water resources. In this case, the new $FWQI_{Lerma}$ is focused in assessing reservoirs located in the same basin, considering different water uses, surrounding land use, and their position into the basin. Like other WQI, their scores are into the range of 0–100, with the superior limit indicating an excellent water quality. Unlike other WQI that use 15–18 water quality variables, $FWQI_{Lerma}$ only uses eight. This is the first WQI that includes SDT as an important parameter, making measurement easier and cost-effective. It is an important issue in a basin with a high number of reservoirs.

Fuzzy inference system has been used by other authors to design WQI for rivers [24, 38], but not in reservoirs or lakes. Liou and Lo [12] applied fuzzy set theory to evaluate trophic state in some reservoirs in China using the three typical parameters: total phosphorous, Chl *a*, and SDT.

Bai et al. [11] and Mourhir et al. [23] proposed a river water quality index based on fuzzy logic, using six indicators and 15,625 and 86 rules, respectively. The fuzzy WQI proposed by [38] was set up with 27 water quality indicators (WQInd) and 96 fuzzy inference rules; while those proposed by [24] is composed of 9 variables and 3125 fuzzy rules. In this study, a multivariate analysis of discriminants and other statistic tools were employed to characterize the reservoirs and select the most important water quality variables; in this sense, $FWQI_{Lerma}$ was set up by eight water quality indicators and 633 inference rules.

While other authors have applied only one of the traditional methods of defuzzification: Centre of Gravity [21], Centroid [11, 21, 38], or MOM [41], this study analysed what could be the best defuzzification method, considering a benchmark. Thus, the best method for defuzzification was LOM. In this sense, comparison with other WQI as a benchmark was a process to know the range of water quality at which the reservoirs should be. In this study, we look for the match with the WQI_{Dinius}.

Ocampo-Duque et al. [38] compared their FWQI with some impact indicators such as biochemical responses in fish, which matched with FWQI spatial data. Semiromi et al. [21] compared their FWQI with other indices using a set of independent data. In this study, $FWQI_{Lerma}$ was compared to WQI_{Dinius} to verify the range of scores and to select the best method of defuzzification, which is a part of the validation process. On the other hand, a set of other reservoirs into the Río Lerma basin was used to evaluate the applicability of this index in other water bodies whose water quality data were not used in the setting, that is, its potential use at the regional or basin levels was tested.

Thus, $FWQI_{Lerma}$ scores were compatible with the water quality status assessed with the WQI_{Dinius}. This index showed that those reservoirs exposed to minimum impact (U, LC, TF, and TP) obtained the best scores, while those reservoirs closed to urban, industrial, or agricultural zones (W, TX, CO) displayed scores with a regular water quality.

6. Concluding remarks

- Water Quality Indices are important tools to assess the status of water bodies considering the integrated measure of physical and chemical indicators that contribute to decision making.
- In concordance with other authors, FL and FIS in this study resulted to be excellent tools to assess the water quality in water bodies.
- $FWQI_{Lerma}$ shows to be consistent with WQI_{Dinius}.
- $FWQI_{Lerma}$ proved to be an outstanding and robust tool to rate and take decisions about the water quality in reservoirs located in the Río Lerma basin since it reflected the water quality scores in the same range as other indices.

http://dx.doi.org/10.5772/64265

- This index is believed to be cheaper because it uses only eight parameters, among them DO and conductivity are measured in the field using a probe (a very common equipment for water quality monitoring), and SDT is recorded with a single Secchi disk.
- This index is one of the WQI specially configured to assess water quality in reservoirs.

Acknowledgements

Authors are thankful to Instituto Politécnico Nacional and CONACyT-SEMARNAT for financial support for the Project "Diagnóstico del estado trófico y calidad del agua de embalses de la cuenca del río Lerma, generación de un modelo regional. C01-0384." (Assessment of trophic state and water quality of reservoirs of the Río Lerma Basin, generating a regional model.). Likewise, authors would like to thank to Ing. Mali Carol Uspango Becerro by editing the figures.

Author details

Jacinto Elías Sedeño-Díaz[1*] and Eugenia López-López[2]

*Address all correspondence to: jsedeno@ipn.mx

1 Polytechnical Coordination for Sustainability, National Polytechnic Institute. Coordinación Politécnica para la Sustentabilidad, Instituto Politécnico Nacional. Av. IPN esq. Wilfrido Massieu s/n, Edif. Biblioteca Nacional 3er piso., Ciudad de México, México

2 Laboratory of Aquatic Ecosystems Health Assessment, National School of Biological Sciences, National Polytechnic Institute. Laboratorio de Evaluación de la Salud de los Ecosistemas Acuáticos, Departamento de Zoología. Escuela Nacional de Ciencias Biológicas, Instituto Politécnico Nacional. Prol. de Carpio y Plan de Ayala, Col. Sto. Tomas, Ciudad de México. México

References

[1] Debels P, Figueroa R, Urrutia R, Barra R and Niell X. Evaluation of water quality in the Chillán River (Central Chile) using physicochemical parameters and a modified Water Quality Index. Environmental Monitoring and Assessment . 2005;110:301–322. DOI: 10.1007/s10661-005-8064-1

[2] Sedeño-Díaz JE, López-López E. Water quality in the Río Lerma, Mexico: An overview of the last quarter of the twentieth century. Water Resources Management . 2007;21:1797–1812. DOI: 10.1007/s11269-006-9128-x

[3] Štambuk-Giljanović N. Water quality evaluation by index in Dalmatia. Water Research. 1999;33(16):3423–3440. DOI: 10.1016/S0043-1354(99)00063-9

[4] Thanh NC, Biswas AK, (editors). Environmentally-sound Water Management. 1st ed. Oxford: Oxford University Press; 1990183 p.

[5] Brown RM, McClelland NI, Deininger RA, Tozer RG. A water quality index: Do we dare? Water and Sewage Works. 1970;117:339–343.

[6] Dinius SH. Design of an index of water quality. Water Resources Bulletin. 1987;23(5): 833–843 .

[7] Walski TM, Parker FL. Consumer's water quality index. Journal of the Environmental Engineering Division. 1974;100(EE3):593–611.

[8] Miller WW, Joung HM, Mahannah CN, Garrett JR. Identification of water quality differences in Nevada through index application. Journal of Environmental Quality. 1986;15(3):265–272.

[9] Bordalo AA, Nilsumranchit W, Chalermwat K. Water quality and uses of the Bangpakong River (eastern Thailand). Water Research. 2001;35(15):3635–3642.

[10] Shultz MT. A critique of EPA's index of watershed indicators. Journal of Environmental Management. 2001;62:429–442.

[11] Bai RV, Bouwmeester R, Mohan S. Fuzzy logic water quality index and importance of water quality parameters. Air, Soil and Water Research. 2009;2:51–59.

[12] Liou Y, Lo S. A fuzzy index model for trophic status evaluation of reservoir waters. Water Research. 2005;39:1415–1423. DOI: 10.1016/j.watres.2005.01.014

[13] Monteiro Masalskiene Roveda SR, Maia Bondança AP, Soares Silva JG, Roveda JAF. Development of a water quality index using a fuzzy logic: A case study for the Sorocaba river. In: IEEE, editor. 2010 IEEE International Conference on Fuzzy Systems (FUZZ); Fuzzy Systems (FUZZ); Barcelona, Spain: 2010. 18–23 July, 2010; p. 1–5. DOI: 10.1109/ FUZZY.2010.5584172[13]

[14] Horton RK. An index number system for rating water quality. Journal of Water Pollution Control Federation. 1965;37(3):300–305.

[15] Pesce SF, Wunderlin DA. Use of water quality indices to verify the impact of Córdoba City (Argentina) on Suquía River. Water Research. 2000;34:2915–2926. DOI: 10.1016/ S0043-1354(00)00036-1

[16] Jonnalagadda SB, Mhere G. Water quality of the Odzi river in the eastern highlands of Zimbabwe. Water Reseach. 2001;35(10):2371–2376.

[17] Sargaonkar A, Deshpande V. Development of an overall index of pollution for surface water based on a general classification scheme in Indian Context. Environmental Monitoring and Assessment. 2003;89(1):43–67.

[18] Liou SM, Lo SL, Wang SH. A generalized water quality index for Taiwan. Environmental Monitoring and Assessment. 2004;96:32 35.

[19] Said A, Stevens DK, Sehlke G. An innovative index for evaluating water quality in streams. Environmental Management. 2004;34(3):406–414.

[20] Kannel PR, Lee S, Lee Y, Kanel SR, Khan SP. Application of water quality indices and dissolved oxygen as indicators for river water classification and urban impact assessment. Environmental Monitoring and Assessment. 2007;132(1):93–110.

[21] Semiromi FB, Hassani AH, Torabian A, Karbassi AR, Hosseinzadeh Lotfi F. Water quality index development using fuzzy logic: A case study of the Karoon River of Iran. African Journal of Biotechnology. 2011;10(50):10125–10133. DOI: 10.5897/AJB11.1608.

[22] Ocampo-Duque W, Osorio C, Piamba C, Shumahmacher M, Domingo JL. Water quality analysis in rivers with non-parametric probability distributions and fuzzy inference systems: Application to the Cauca River, Colombia. Environment International. 2013;52:17–28.

[23] Mourhir A, Rachidi T, Karim M. River water quality index for Morocco using a fuzzy inference system. Environmental Systems Research. 2014;3(21):1–12.

[24] Lermotov A, Yokoyama L, Solares A. River quality analysis using fuzzy water quality index: Ribeira do Iguape river watershed, Brazil. Ecological Indicators. 2009;9:1118–1197.

[25] Sahu M, Mahapatra SS, Sahu HB, Patel RK. Prediction of water quality index using neuro fuzzy inference system. Water Quality, Exposure and Health. 2011;3:175–191. DOI: 10.1007/s12403-011-0054-7.

[26] Bernal-Brooks FW, Dávalos-Lind L, Lind OT. Seasonal and spatial variation in algal growth potential and growth-limiting nutrients in a shallow endorheic lake: Lake Pátzcuaro (Mexico). Lakes Reservoirs: Research and Management. 2003;8:83–93. DOI: 10.1046/j.1320-5331.2003.00217.x.

[27] López-López, E, Soto-Galera E. Diagnosis of eutrophication of the reservoir Ignacio Allende, Guanajuato, Mexico. Annals of the Institute of Marine Sciences and Limnology. 1993;20(1):33–42.

[28] Lopez E. Regional limnology of ten reservoirs in the Lerma Basin, Mexico. International Association for Theoretical and Applied Limnology. 2000;274:2288–2293.

[29] Xiangcan J. Analysis of eutrophication state and trend for lakes in China. Journal of Limnolgy. 2003;62(2):60–66. DOI: http://dx.doi.org/10.4081/jlimnol.2003.s1.60.

[30] Cotler-Ávalos H. Water Quality of Reservoirs. In: Cotler-Ávalos E, Mazari Hiriart M, and De Anda Sánchez J. (editors). Atlas of the Lerma-Chapala Basin, Building a Joint

Vision, Edition: 1, Secretary of the Environment and Natural Resources, National Ecology Institute and National Autonomous University of Máxico, 2006; pp. 148–150

[31] Shuhaimi-Othman M, Lim EC, Mushrifah I. Water quality changes in Chini Lake, Pahang, West Malaysia. Environmental Monitoring and Assessment. 2007;131:279–292. DOI 10.1007/s10661-006-9475-3.

[32] Lu RS, Lo SL, Hu JY. Analysis of reservoir water quality using fuzzy synthetic evaluation. Stochastic Environmental Research and Risk Assessment. 1999;13:327–336.

[33] Liou YT, Lo SL. A fuzzy index model for trophic status evaluation of reservoir waters. Water Research. 2005;39(7):1415–1423. DOI:10.1016/j.watres.2005.01.014.

[34] Lu RS, Lo SL. Diagnosing reservoir water quality using self-organizing maps and fuzzy theory. Water Research. 2002;36(9):2265–2274.

[35] Cotler-Ávalos H. Dams and levees. In: Cotler-Ávalos E, Mazari Hiriart M, and De Anda Sánchez J. (editors). Atlas of the Lerma-Chapala Basin, Building a Joint Vision, Edition: 1, Secretary of the Environment and Natural Resources, National Ecology Institute and National Autonomous University of México, 2006; pp. 148–150.

[36] Zadeh LA. Fuzzy sets. Information and Control. 1965;8:338–353.

[37] Sivert W. Ecological impact classification with fuzzy sets. Ecologial Modelling. 1997;96:1–10.

[38] Ocampo-Duque W, Ferré HN, Domingo J, Schuhmancher M. Assessing wáter quality in rivers with fuzzy inference systems: A case study. Environmental International. 2006;32:733–742.

[39] Castillo O, Melin P. Type-2 Fuzzy Logic Theory and Applications. Berlin: Springer-Verlag; 2008.

[40] Dahiya S, Singh B, Gaur S, Garg VK, Kushwaha HS. Analysis of groundwater quality using fuzzy synthetic evaluation. Journal of Hazardous Materials. 2007;147:938–946.

[41] Altunkaynak A, Sen Z. Fuzzy logic model of lake water level fluctuations in Lake Van, Turkey. Theoretical and Applied Climatology. 2007;90:227–233. DOI: 10.1007/s00704-006-0267-z.

[42] Kaurish FW, Younos T. Developing a standardized water quality index for evaluating surface water quality. Journal of the American Water Resources Association. 2007;43(2): 533–545.

Chapter 10

Water Quality Modelling of Northern Lakes Case Study (Egyptian Northern Lakes)

Noha Donia

Additional information is available at the end of the chapter

http://dx.doi.org/10.5772/63526

Abstract

Since the shallow coastal lakes are not only one of the most valuable ecosystems in the world but also some of the most threatened as they receive the wastewater discharged from the watershed, it was important to develop a more detailed modelling component for the lake system. Nowadays, relative to the present advances in computational sciences, hardware and software, improvement in rivers, catchments and lakes modelling has been only modest since the past few decades. The main objective of the study is to examine and evaluate the impact of alternative water quality management practices in the selected drainage catchment, and their effect on the environmental condition of the lake as an important component of the watershed. A hydrodynamic and water quality model was used to study the current status of coastal lakes subject to the discharges and pollution loadings coming from the agricultural drains and the point sources discharge directly to the lake, through simulating the flow circulation inside the main basin of the lake, the transport and advection of the pollutants due to the effluent discharges from drains and other sources of pollutants, and identify and develop the most critical surface drainage water quality indicators to simulate and predict the temporal and spatial variation of pollution.

Keywords: water quality, modelling, coastal lakes, pollution, Mariout Lake

1. Introduction

The aim of modelling of surface water quality is to construct a mathematical model of the water body in order to simulate variation in water quality with the variation in initial and boundary conditions. The modelling is applied to solve problems related to water quality by analysing the occurring phenomena and finding dependencies between them as well as to attempt to predict and quantify effects of the changes in the aquatic environment [1]. Over the past 75 years, engineers have developed water quality models to simulate a wide variety of pollutants in a broad range of receiving waters. In recent years, these receiving water models are being coupled with models of watersheds, groundwater, bottom sediments to provide comprehensive frameworks predicting the impact of human activities on water quality [2].

Mathematical modelling of lakes water quality started to receive high attention in the 1960s. According to [3], mathematical models of lakes have evolved along two different lines. First, there was the extension of the zero-dimensional model to one-, two- and three-dimensional models. Then, there were the modelling activities that focused primarily on a better and more detailed description of the chemical-biological processes [4]. From the survey of the literature, many lake models have been applied in various regions, and as a result of several applications, models have become more and more complex.

Modelling of water quality in lakes involves the representations of effluent quality, mixing pattern, physical and chemical processes and biological growths and their role in the removal and release of substances. Such models can be classified into physical, chemical or biological models that simulate lakes eutrophication. Another classification may be as long-term planning models or short-term operational models. Several water quality studies have been performed on lakes in general and on shallow lakes in particular, for example, [5–9].

The objective of this chapter is to illustrate the technique of building a hydrodynamic and water quality model as application on one of the Egyptian coastal lakes (Mariout Lake).

2. Lake Mariout, one of Egyptian Northern Delta lakes

The northern delta lakes provide many economic, environmental and social benefits to the people of Egypt and the Mediterranean. Some of these benefits are easy to quantify. For example, the 1998 catch from the four lakes — Burullus, Idku, Manzala and Mariout — amounted to LE 1.05 billion, or roughly 35 percent of the country's total fish income. The lakes currently provide passive primary and secondary treatment of wastewater that would be equivalent to hundreds of millions of dollars worth of new treatment plants. Other important and valuable benefits are much harder to quantify. It is unknown how much property damage and economic dislocation Lake Mariout prevented in 1992 when Alexandria experienced severe flooding, or how much the lake contributes to agricultural production by buffering against seawater intrusion of groundwater supplies. Beyond Egypt, it is difficult to value the benefit that those wetlands provide to sustain migratory birds of the entire Eastern Mediterranean/Black Sea region. In the future, with predicted sea level rise and the frequency of coastal storms on the increase, the lakes may be even more important to prevent natural disasters [10].

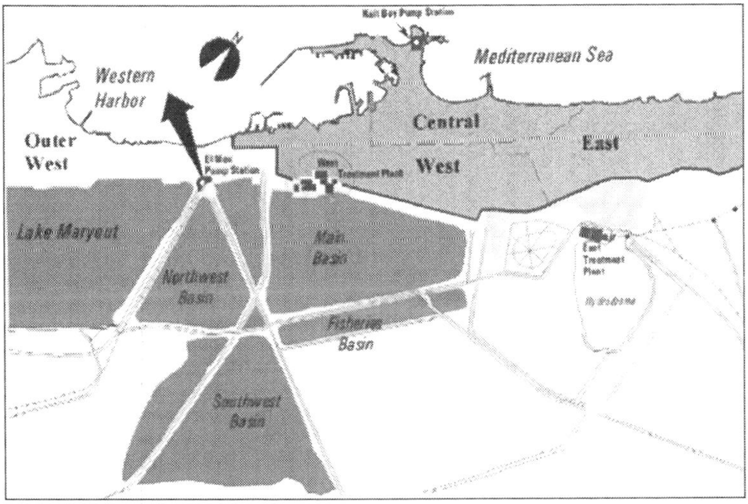

Figure 1. Lake Mariout location.

Figure 2. Lake Mariout satellite image and sample location.

Lake Mariout is the smallest of the northern lakes and perhaps the most threatened. Lake Mariout lies between Latitude 31° 07′ N and Longitude 29° 57′ E along the Mediterranean coast of Egypt. The lake environment was continuously subjected to quality degradation due to human pressure as well as land reclamation reducing the area of the lake.

Currently, the lake is divided artificially into four main basins as shown in **Figures 1** and **2**, namely, 6000 feddans basin (Main Basin), 5000 feddans basin (South Basin), 3000 feddans basin (West Basin) and 1000 feddans basin (Aquaculture Basin). These ponds are dissected by roads and embankments as follows [11]:

- The Main Basin is about 14.77 km^2 with an average depth of 0.8 m. This basin receives water from the El-Nubariya canal and Omoum drain, the heavily polluted water by industrial wastes; and untreated sewage from municipal and industrial outfalls of El-Qalaa drain had been diverted through the new Richa drain. West Wastewater Treatment Plant effluent had been discharged along the north of the basin. One minor inflow is a discharge of waste from a textile plant into a ditch which crossed Qabarry. The Main Basin is bisected by the Nubariya canal, and the triangular area between this canal and the Omoum drain is also considered as part of the Main Basin.
  - The Western Basin is about 11.59 km^2 with average water depths of 0.7 m. Adjacent to this basin, salt marshes are located and are producing 1,000,000 kg of unrefined salt per year. They are surrounded by many industrial and petrochemical companies.
  - The Southern Basin covers 33.77 km^2, and is partially divided by El-Nubariya canal, although breaks in the canal embankments allow water to pass from one sub-basin to the other. This basin is very shallow and average water depths are 0.68 m. The main source of water is El-Omoum drain and El-Nubariya canal. Along the length of the El-Omoum, a series of breaches allow flow to leave the drain and enter the basin. Along the western boundary, a series of breaches allow exchange of water between the basin and the El-Nubariya canal. This basin consists of heavily vegetated areas and fish farms. Also, considerable wetland loss in this portion of the basin was recorded. Many petrochemical and petroleum companies, such as Amria and Misr Petroleum companies, discharge their wastes into the north part of this basin.
  - The (Fisheries) Aquaculture Basin covers 9.44 km^2 (849 feddans), and it consists of a series of small basins separated by earthen berms. This facility is a research centre for fish farming and is operated by the Alexandria Governorate. There are two sources of water for this facility. One is small pump stations which pump 400,000 m^3/day from Abis drain and which run parallel to the basin. The other is small openings from El-Omoum drain.

  Comparison of the chemical composition of Lake Mariout water with that of proper sea water and drainage water shows that the lake water presents an intermediate composition between both sea and drainage water. This phenomenon can be explained by seepage from the sea. Such explantation is supported by the low level of water and by the water balance which is supported by older data of the salt content of wells in Mariout region. There are three main canals (El-Qalaa, El-Omoum and El-Nubariya) that are considered the main inflows to the lake. El-Qalaa drain is located at north-east while El-Omoum and

http://dx.doi.org/10.5772/63526

El-Nubariya canals are at the east and south of the lake, respectively. Other inflows are the water treatment plant (WTP) and the discharges from the petrochemical area nearby the north western basin. El-Omoum and El-Nubariya canals are less polluted drains, considering their nutrients (N, P) and DO concentrations. El-Omoum receives mainly agricultural drainage water; moreover, the drain receives both raw and treated wastewater from several defined and undefined sources. Therefore, these drains also contribute to the nutrient loadings in the lake, but to a lesser extent. Additionally, non-point sources such as agricultural run-off containing pesticides and fertilizers are also contributing to the deterioration of the environmental quality of the lake [12].

As a result of the high nutrient loading, the lake has become anthropogenic-polluted and eutrophic. Eutrophication of lakes is a natural process that can be accelerated by man's activities that introduce an excess of nutrients together with other pollutants. Main sources of nutrients and pollutants can include: human sewage, industrial waste, farm and urban run-off. Currently, the lake is 60% covered by aquatic vegetation (*Phragmites australis* and *Eichornia crassipes*). High nutrients and low DO concentrations have been observed specially at the Main Basin, which in turn affects the ecological processes occurring in the lake and therefore its whole environmental condition [13].

Applying a hydrodynamic and water quality numerical modelling study at Lake Mariout will help to give some answers to both planning and technical questions of water quality managers, decision makers and those of technical engineers working on the sampling, monitoring and analysis of water quality parameters. Specifically, the main objectives of the hydrodynamic and water quality numerical model study can be summarised as follows:

- Studying the current status of the Lake Mariout and using the available data to simulate the Main Basin of Lake Mariout subject to the discharges and pollution loadings coming from the agricultural drains and the point sources discharge directly to the lake.
- Investigate the flow circulation inside the Main Basin of Lake Mariout and its effect in minimizing the negative impacts on the water quality of the lake.
- Investigate the transport and advection of the pollutants due to the effluent discharges from drains and other sources of pollutants.

- Identify and develop the most critical surface drainage water quality indicators to simulate and predict the temporal and spatial variation of pollution.
- Examine and evaluate different modelling scenarios to study the impact of alternative water quality management practices in the selected drainage catchment, and their effects on the environmental condition of the lake as an important component of the watershed.
- Perform sensitivity analysis for modelling parameters and variables, showing the response of the model to influential parameters and coefficients used in the modelling process, especially those with high degree of uncertainties on their values.
- To achieve the study objectives, the following scope of work can be summarised as follows:

- Data collection including both hydrographic and bathymetric survey for the Main Basin of Lake Mariout, which is necessary to fulfil the hydrodynamic and water quality simulations of the numerical model of the main basin of Lake Mariout.
- Develop a two-dimensional hydrodynamic and water quality numerical flow model to simulate the flow pattern in the lake vicinity of the study area, and the discharges and pollution loadings coming from the agricultural drains and the point sources discharge directly to the lake.
- After the model development, calibration is conducted in order that the model will be ready for different potential model scenarios. This will help to investigate the impact of alternative water quality management processes and their effects on the environmental condition of the lake. The analysis of the model scenarios forms the basis to assess and select the optimum solution for minimizing the pollution coming from the agriculture drains and other point source of pollution to the lake.

3. Data collection and field measurements

The setup, testing and application of a lake model of hydrodynamics and water quality require a variety of different data sets to specify boundary or input conditions and for model calibration and verification.

In case of Mariout Lake, data collection includes historical data on the wind conditions, water temperature, evaporation rate and the precipitation rate in the project site. Wind data were extracted from the work of [14], the data show that the predominant wind direction is 22.5° NW with a wind speed of approximately 3.75 m/s. The average monthly temperature in Lake Mariout ranged from 13 to 29°C in a study carried by [15]. The annual average evapotranspiration used in the model was calculated with the Penman-Monteith method [16], where the crop coefficient *Kc* (reed) used in the calculation was extracted from a study based on field experiment and measurements carried out in three locations in the UK [17]. The precipitation value used in the model corresponds to the average precipitation of year 2007, (0.66 mm/day) as presented in the Lake Mariout data acquisition report (NIOF, 2007[sn1]). The topography and bathymetry data used in the model were provided by NIOF in a DEM format with resolution of 45 m reference is made to [18].

Field measurements were carried out in coordination with the National Institute of Oceanography and Fisheries (NIOF) for 2 weeks. The samples were taken from nine sites representing the Main Basin and discharge points of Qalaa drain, Omoum drain, Nubariya canal and El-Max pumping station as shown in **Figure 2**; the measurements comprised the following:

- Water flows (m^3/h) which determines the inflow, outflow in the Main Basin.
- Water levels within the basin to a fixed point.
- Basic physical parameters: temperature, salinity and total suspended matter.

- Organic matter of the lake.
- Nutrient variables: ammonia, nitrates and phosphorus compounds.
- Biological data including: chlorophyll-a, phytoplankton, zooplankton.
- Microbiological data: faecal coliform and total coliform.

Results of field measurements of hydraulic parameters are shown in **Table 1** and results of field measurements of water quality parameters are shown in **Table 2**.

Site no.	Site name	Cross section (m^2)	Average weekly water velocities (m/s)	Average weekly water discharges (m^3/hour)
2	Nubariya canal (desert road)	122.00	0.44	192480
3	Omoum Drain (desert road)	097.00	0.42	146658
5	Fisheries Hole in dam	001.60	0.30	001699
6	End Omoum diversion before Nubaria	136.00	0.11	053000
7	El-Max Pumping Station	243.00	0.30	263193
8	Western Water Treatment Plant	003.40	1.22	015105
9	Qalaa Drain outlet in Main Basin	008.90	0.86	027602

Table 1. Water flow measurements.

Code	Temp °C	Trans cm	EC mS/cm	TDS g/l	TSS g/l	Sal ‰	pH	DO mg/l	BOD mg/l	COD mg/l	NH3 µg/l	NO2 µg/l	NO3 µg/l	TN µg/l	PO4 µg/l	Tp µg/l
West Nubaria PS	25.7	35	7.68	4.82	0.040	4.81	7.18	7.38	4.90	22.09	971	89.6	332.6	1981.2	54.2	115.36
Nubaria Canal Desert Road	22.1	60	5.26	3.49	0.034	3.48	7.68	5.60	4.12	21.64	723	106.4	529.4	1844.7	66.7	161.04
Ommoun Desert road	22.3	60	3.45	2.34	0.034	2.33	7.52	5.82	4.66	20.22	2164	150.5	470.1	3492.6	194.7	566.64
End of Qalaa Diversion Canal before Nubaria Canal	22.8	15	2.39	1.21	0.111	1.21	7.33	0.00	111.56	88.96	19956	0.00	0	22856.1	915.2	1203.84
Western WTP	24.6	10	1.99	1.14	0.125	1.14	7.12	0.00	140.12	92.92	20996	0.00	0	24869.3	1019.7	1335.84
Main Basin	21.1	40	3.65	2.40	0.034	2.39	8.68	9.12	5.06	32.32	2640	102	150.8	3886.>	165.8	436.92
Noha El Max station	21.3	35	5.9	3.59	0.043	3.58	7.32	5.42	4.92	40.84	4610	92.6	212.5	6365.7	190.3	528.24
Fisheries hole in dam	22.1	15	2.41	1.28	0.115	1.28	7.67	0.00	120.04	90.69	19670	0.00	0	22886.4	928.4	1244.76
End of Qalaa drain	23.8	20	2.34	1.26	0.109	1.26	7.22	0.00	123.12	89.50	20030	0.00	0	23386.1	905.3	1236.84

Table 2. Measured water quality parameters in Mariout Lake.

The following section represents the hydrodynamic and water quality modelling studies that were carried out to investigate the efficiency of the water circulation system and water quality parameters inside the Main Basin of Mariout Lake. The model setup, calibration and the analyses of the results of model scenarios are included. Depending on the model results and analysis, the conclusions and recommendations are presented.

4. Water quality model development

Delft3D Software Package of Delft Hydraulics, the Netherlands, was used to develop the hydrodynamic numerical flow and water quality model which simulates the flow pattern and the water quality inside the lake. Delft3D is a integrated, powerful and flexible software, which was developed by Deltares, the Netherlands. The hydrodynamic and water quality modules were used in this study. Consequently, a brief explanation of these modules is in the following section.

The FLOW module of Delft3D is basically a multi-dimensional (2D and 3D) hydrodynamic (and transport) simulation which calculates non-steady flow and transport phenomena resulting from tidal and meteorological forcing on a curvilinear, boundary-fitted grid [19]. The WAQ module of Delft3D for water quality modelling the spatial resolution generally consists of the resolution of the underlying flow field as generated by the hydrodynamic model itself or of flows on integer multiples of those hydrodynamic grid cells. For water quality modelling, there also is external forcing in the form of waste loads, meteorology, open boundary concentrations, etc. [20].

4.1. Setup of the hydrodynamic model

The hydrodynamic model simulates the flow pattern in the Main Basin vicinity. All parameters and variables in the model have units according to the SI conventions. The coordinate system used for the model is in WGS-84 Geographic UTM system Zone 35. All metric coordinates in this report will be given in this coordinate system. The depths and water level information in the flow model are defined relative to a levelling datum, which is equal to mean sea level (MSL). The following sections present the steps of development of Mariout hydrodynamic model.

4.1.1. Grid generation

The first step in the schematization process is the design and generation of the computational grid. The computational grid is a curvilinear grid to avoid the stair case problem which affects the numerical accuracy. In the design of a curvilinear grid, it is important to follow the land boundaries as good as possible. For the generation of a computational grid, the following items are important:

- the areas which require the highest resolution;
- the orthogonality of individual cells;

http://dx.doi.org/10.5772/63526

- the spatial variation of the dimensions of the cells;
- the total number of computational points.

The resulting computational grid is a compromise between the above items, the selected dimensions of the model and the location of the boundaries. The general layout of the computational grid of the Mariout model is given in **Figure 2**.

4.1.2. Depth schematization

The schematization of the land boundaries and the water depths have been derived from the hydrographic survey data. The bathymetric data have been mapped through an interpolation procedure on the computational grid of Mariout model. In this way, each coordinate of the computational grid of the model is given a depth value. The transition between the regions covered by different bathymetric data sources have been checked and smoothed where necessary.

4.1.3. Boundary conditions

In the flow simulations of a specific area with two open boundaries, it is preferable to set up one boundary as a discharge boundary and the other one as a water level boundary. In Lake Mariout model, the open boundary for Nubariya Canal and Omoum Drain were selected as a discharge boundary. The boundary at El-Max Pumping station was selected as water level boundary, while other sources like Qalaa drain and the West Water Treatment Plant were

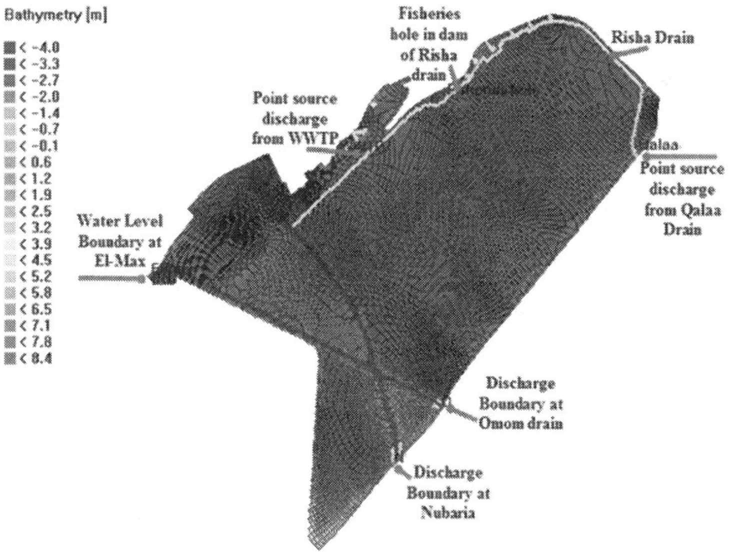

Figure 3. Model schematization with all boundaries and sources of discharge.

simulated as source point discharge. During the calibration phase, the discharges and water level measurements at the location of the open boundaries and at the other sources of water were used in the model. In the model scenarios (production simulations), the discharge data imposed in the discharge boundary is based on the dominant flow condition. The water levels associated with these discharges were used for each scenario as a water level boundary. The relevant water levels associated to these discharges were obtained from the historical data available about the Lake Mariout. **Figure 3** shows the model schematization with all boundaries and the source points of discharges.

4.1.4. Parameter settings

A uniform water density of 1025 kg/m^3 was used, representing the salt water density. The acceleration of gravity was set to 9.81m/s^2. The value for the horizontal eddy viscosity is set to 1.0 m^2/s. The time step was selected for the model simulations based on the grid size and the Courant Number. Time step of 0.5 min (30 s) was used in the simulations. This time step fulfils the numerical criteria and the Courant Number requirements.

4.1.5. Model calibration

During the model calibration, the measured depth averaged flow velocities and water levels which were carried out by the National Institute of Oceanography and Fisheries (NIOF) were compared with the model results. Tuning of the roughness parameter in the model was carried

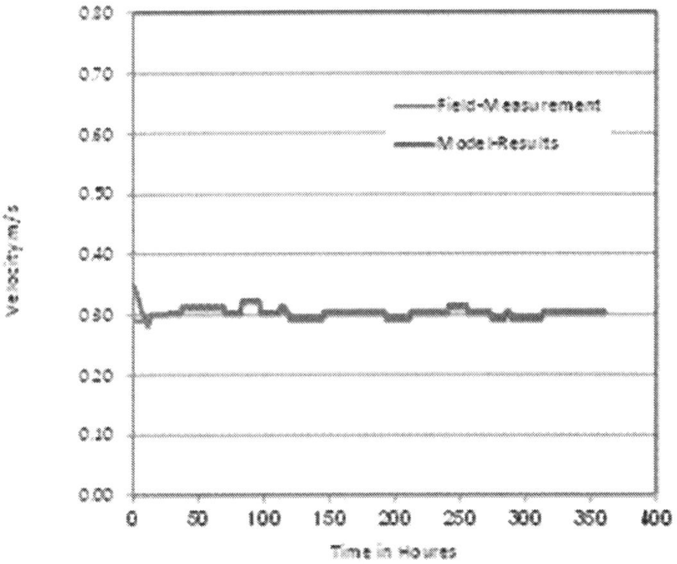

Figure 4. Flow velocity comparison at point fisheries hole in dam.

http://dx.doi.org/10.5772/63526

out to obtain the best match between the model and the field measurements. Manning roughness coefficient was varied between 0.02 at non-vegetated area and 0.06 at the heavy vegetated area along the model area to give the best match between the measurements and the model computations. **Figure 4** shows the comparison between the measured and computed flow velocity values. The results for water level and currents were in good agreement with the measurements, which confirms that the model simulates the flow pattern in the main basin of Lake Mariout in the right way.

4.2. Water quality model setup

To apply the Delft3D-WAQ module, the following steps must be followed:

- Get the result from the hydrodynamic simulation and make it suitable for application in the water quality simulation (coupling process).
- Selection of the substances and water quality processes to be included in the model.
- Preparation of initial conditions, boundary conditions, waste loads, simulation time, output variables and identification of monitoring points.
- Run the simulation and check the output.
- Calibrate and verify the model.

4.2.1. Selection of the processes involved in the water quality model

In Delft3D-WAQ module, the constituents of a water system are divided into functional groups. A functional group includes one or more substances that display similar physical and/or (bio)chemical behaviour in a water system. Functional groups can interact with each other directly or indirectly. PLCT (Processes Library Configuration Tool) is used to choose the substances and water quality processes to be modelled. The selected substances groups and parameters are described in **Table 3**.

Substance group	Selected model parameters	Associated processes
General	Continuity, water temperature, salinity	Temperature and heat exchange
Oxygen-BOD	BOD-COD-DO	Mineralization BOD and COD, sedimentation COD, re-aeration of oxygen
Suspended matter	Inorganic matter (TSM)	Sedimentation, resuspension
Eutrophication	Ammonium (NH4), nitrate (NH3), ortho-phosphate (PO4)	Nitrification of ammonium Denitrification of nitrates

Table 3. Model parameters and associated processes.

4.2.2. Model boundary conditions and observation locations

Average historical monthly values have been selected for initial conditions of water quality parameters inside the lake. The continuity parameter which checks the mass balance of the model was set to 1 g/m^3. The model simulation period was selected as the same period for the hydrodynamic modelling, namely, for 1 month. Water quality model time step was set to 1 min. The default values were taken as input for some selected modelled substances, that they are by default constant in time and space. However, process parameters are changed in the process parameters data group because they can vary in time and/or space. Initially, process parameters will have the default value that is taken from the PLCT.

The water quality model boundary sections are selected to be the same boundary sections for the hydrodynamic model at the locations of the main input sources to the lake, where all discharges enter the lake shown in **Figure 3**. At the two sections for the Omoum drain outlet and Nubariya canal outlet, concentrations for different modelled parameters are defined as time-varying boundary conditions. The concentrations used at the boundaries are time series average monthly concentrations for the modelling period.

4.2.3. The water quality model calibration

Figure 5 shows the simulation of dissolved oxygen in Mariout Lake as an example of output from the model. In this study, the water quality model calibration is done on the conventional water quality parameters or oxygen group, nutrients group and coliform group (faecal and total) and process parameters are adjusted for of calibration. The model calibration was carried

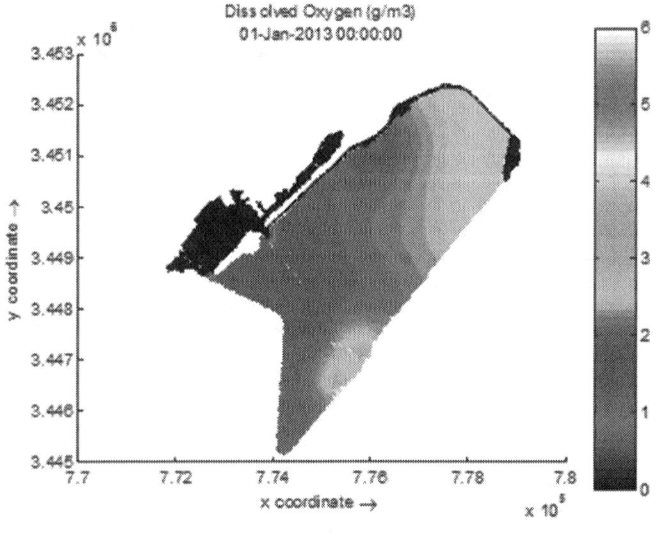

Figure 5. Simulation of dissolved oxygen in Lake Mariout.

http://dx.doi.org/10.5772/63526

out by visual comparison of simulations and measurements in graphs, together with the calculation of the statistical error values such as mean relative error (MRE), the root mean square error (RMSE) to examine the performance of the model.

The simulated water quality parameters were plotted in graphs to make comparisons with respect to the observations in Lake Mariout during field survey, which were used to check how the simulations fit the observations. Besides, MRE was used to quantify the agreement of the model, by dividing the residuals by the observed values. In this study, the calculation of RE and MRE was based on the following equations:

$$\mathrm{RE}=(\mathrm{Csim}\text{-}Cobs)\times 100 \div Cobs$$

$$Sum\, RE \div n$$

where Csim and Cobs are the simulated and observed values, respectively, and n is the number of cases. The MRE denotes the mean relative difference between simulations and observations.

Table 4 shows the different values of RE and MRE for the modelled parameters at this level. **Figure 6** shows the calibration results of dissolved oxygen that shows good agreement with field measurements It is noted that at the entrance of the Qalaa drain to the lake, the DO has the lowest values; in general, the DO measurements are close to the simulated results with an RME value of 5.11%. **Figure 6** shows the calibration results of dissolved oxygen that shows good agreement with field measurements.

Location/parameter	DetN (RE%)	NH4 (RE%)	N03 (RE%)	CBOD5 (RE%	COD (RE%)	DO (RE%)	FCOLI (RE%)	TCOLI (RE%)
Omoum Drain	5.13	5.87	1.70	3.41	1.84	0.40	27.78	0.64
WWTP	10.98	8.96	14.63	13.23	10.35	9	19.31	0.07
Fisheries Hole	3.36	1.36	6.54	1.69	4.00	9	10.14	1.11
Elmax	1.88	9.03	3.36	2.97	7.30	7.7	0.03	9.77
Main Basin Middle	0.09	8.46	6.87	6.93	11.51	0	2.39	3.22
Nobariya Canal	0.75	2.69	0.52	2.69	0.04	0.70	6.09	1.18
Qalaa Drain	10.89	4.57	6.87	4.08	4.19	9	0.41	13.31
MRE	4.72	5.01	5.79	5	5.6	5.11	9.45	4.19

Table 4. Relative error for the calibrated model parameters.

The simulated BOD and COD results are very close to the measured values at most locations within the lake, and the MRE value is around 5% for BOD5 and 5.6% for COD. For eutrophication parameters group, the values show agreement between measured and modelled parameters with mean relative error 4.7% for DetN, 5% for NH4 values and 5.7% for PO4 values that are considered acceptable for this kind of water quality modelling. For bacterial parameters, the faecal coliform values show a difference between simulated and observed values at locations in the Omoum drain outlet station, with a relative error of 27%. This could be due to the low-velocity distributions at these locations around the lake edges; but the overall MRE for all measurement locations is within an acceptable range of 9% for faecal coliform and 4.2% for total coliform.

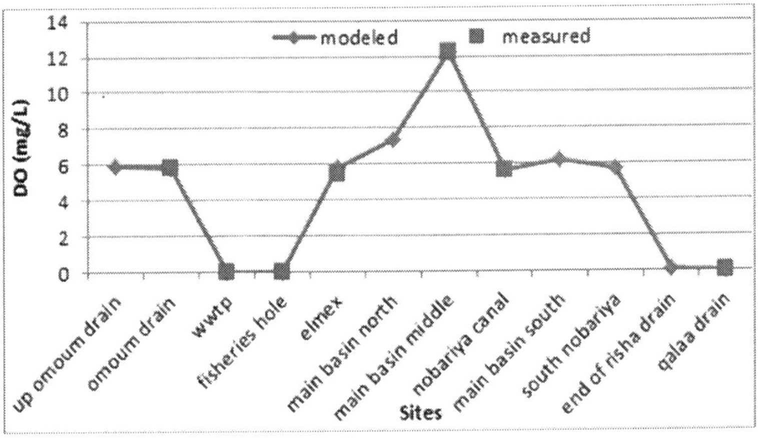

Figure 6. Comparison between measured and modelled DO.

5. Conclusions

Unfortunately, Lake Mariout, one of the Egyptian coastal lakes, suffers from almost all possible environmental problems. In order to evaluate the environmental condition of the Lake Mariout, a 3D hydrodynamic and water quality model that simulates the lake response to pollution loading from the watershed has been developed using the Delft3D hydrodynamic module coupled with the DWAQ module. The model refers to the lake's Main Basin model including watershed simulation scenarios.

First, the 2D hydrodynamic model was developed to simulate the hydrodynamic behaviour of the lake through simulating the water velocity, current and flow within the lake basin. The developed, well-structured hydrodynamic model was also capable of describing the physical and hydrodynamic processes of the water system. Second, a reliable water quality model lake system in this research work is coupled with the developed and calibrated hydrodynamic 2D model. The basic water quality modelling component simulates the main water quality parameters including the oxygen compounds (BOD, COD, DO), nutrients compounds (NH4, TN, TP) and finally the temperature, salinity and inorganic matter.

The calibration was conducted to compare the model results with the observed data at the different locations for both the hydrodynamic and the water quality models. The model results and calculations are in reasonable agreement with the measured concentrations. This developed calibrated model is able to predict the basic water quality indicators of the lake system and ready to conduct any scenarios for watershed water quality management.

Acknowledgements

Special thanks to the Alexandria Coastal Zone Management Project (2010–2015) financed by the Global Environment Facility (GEF) managed through the World Bank in coordination with the Egyptian Environmental Affairs Agency (EEAA) for providing the data required to accomplish this work.

Author details

Noha Donia

Address all correspondence to: ndonia@gmail.com

Institute of Environmental studies and Researches, Ain Shams University, Cairo

References

[1] Chapra S., Surface Water Quality Modeling. 1997; MacGraw Hill, N.Y.

[2] Chapra S. C., Engineering water quality models and TMDLs. Journal of Water Resources Planning and Management, 2003; Vol. 129(4): pp. 245–355.

[3] Jorgensen S. E., Ecological modeling of lakes. In Orlob G.T., Mathematical Modelling of Water Quality: Streams, Lakes and Reservoirs. 1983; John Wiley & Sons, New York, ISBN 047-1100315.

[4] Jorgensen S. E., Kamp-Nielsen L., Christensen T., Windolf-Nielsen J., Westergaard B. Validation of a prognosis based upon a eutrophication model. Ecological Model, 1986; Vol. 32: pp. 165–182.

[5] Collins C. D., Evaluating Water Quality for Lake Management. Final Report. Technical Report, 1988; PB-89-148159/XAB. New York State Museum, Albany, NY, USA.

[6] Stephan G. H., Fang X., Model simulations of dissolved oxygen characteristics of Minnesota lakes: past and future. Environmental Management, 1993; Vol. 18(1): pp. 73–92.

[7] Sagehashi M., Sakoda A., Suzuki, M. A mathematical model of a shallow and eutrophic lake (the Keszthely Basin, Lake Balaton) and simulation of restorative manipulations. Water Research, 2001; Vol. 35(7): pp. 1675–1686.

[8] Buttcher D., Approaches for Nutrient Management in the Lake Okeechobee Watershed, Symposium Handbook, Practical Management for Good Lake Water Quality. 2003; New Zealand.

[9] Zacharias I., Gianni A., Hydrodynamic and dispersion modeling as a tool for restoration of coastal ecosystems. Application to a re-flooded. Environmental Modelling and Software, 2008; Vol. 23(6): pp. 751–767.

[10] EEAA (The Egyptian Environmental Affairs Agency), Annual Report for the Environmental Monitoring Program for the Northern Lakes, 2012; Ministry of Environment, Egypt.

[11] ALAMIM (Alexandria Lake Mariout Integrated Management), Integrated Action Plan, EC-SMAP III, March 2009; American Public Health.

[12] Hossam M. N., Salem A. A. S., Evaluation of drainage water quality for reuse: a case study of the Omoum drain in Egypt. Lowland Technology International, 2003; Vol. 5(2): pp. 27–38.

[13] Mateo M. Á., Lake Mariout: An Ecological Assessment, Laura Serrano and Oscar Serrano (CEAB-CSIC), WADI Project (Water Demand Integration; INCOCT- 2005-015226) and CEDARE (Centre for the Development of the Arabic Region), 2009.

[14] SOGREAH, Alexandria Integrated Coastal Zone Management AICZM, Egyptian Pollution Abatement Project EPAP II,). Base line Conditions, 2008.

[15] Mahlis A. M., El-Wakeel S. K., Morcos S. A., The major cations in Lake Mariout waters. Hydrobiologica, 1970; Vol. 36(2): pp. 253–274.

[16] Allen R. G., Pereira L. S., Raes D., Smith M., Crop Evapotranspiration Guidelines for Computing Crop Water Requirements. FAO Irrigation and Drainage, Paper 56. 1998; FAO, Rome.

[17] Fermor P. M., Gilbert J. C., Gowing D. J. G., Reedbed evapotranspiration rates in England. Hydrological Processes, 2001; Vol. 15(4): pp. 621–631.

[18] NIOF (National Institute of Oceanography and fisheries). Lake Mariout Data Acquisition. 2008.

[19] Delft3D-FLOW, Simulation of Multi-Dimensional Hydrodynamic and Transport Phenomena, Including Sediments. 2015; Delaters, The Netherlands.

[20] D-Water Quality, Water Quality and Aquatic Ecology Modelling Suite. 2015; Delaters, The Netherlands.